SOUNDING THE CENTURY

BILL WAS SO CARELESS OF THE HONOUR OF THE REGIMENT, THAT, HOME ON LEAVE, HE STOOD ON THE TOILET SEAT IN FULL UNIFORM TO FIX THE CISTERN OF THE LAVATORY.

(See page 129)

SOUNDING THE CENTURY

BILL LEADER & CO

---

1 – Glimpses of Far Off Things: 1855-1956

Mike Butler

Illustrated by Peter Seal

A Wang Dang Doodle Production

Matador
9 Priory Business Park,
Wistow Road, Kibworth Beauchamp,
Leicestershire. LE8 0RX
Tel: 0116 279 2299
Email: books@troubador.co.uk
Web: www.troubador.co.uk/matador
Twitter: @matadorbooks

ISBN 978 1800460 768

British Library Cataloguing in Publication Data.
A catalogue record for this book is available from the British Library.

Printed and bound by CPI Group (UK) Ltd, Croydon, CR0 4YY
Typeset in 10pt Minion Pro by Troubador Publishing Ltd, Leicester, UK

Matador is an imprint of Troubador Publishing Ltd

To my quixotic muse, Eva

Contents

Acknowledgements		ix
Foreword		xi
Preface		xiii
Introduction		xxi
I	Gas Works, Palaces and Earthly Damps	1
II	Other Chimney Corners	19
III	Ballyhoo and Trauma (Do You Want Us to Lose the War?)	53
IV	Thistlethwaite, the Road Not Taken	93
V	Hardcastle Hearts	113
VI	Ye Suspects of England	137
VII	In the Dark	159
VIII	William and the Cambridge Spies	179
IX	A Shake to the Business	191
X	Ducking and Diving with Punch and Judy	215
Appendix	Possibly Significant People	235
Index		297

ACKNOWLEDGEMENTS

NONE OF THIS WOULD be possible without an angel; he knows who I mean (he performs acts of gallantry with 78s as Victor Swanvesta). Alistair Banfield, a discographer with a deadly aim, Chris Ackroyd, a gentleman, and Peter Seal, a true artist, went that extra mile. John Foreman was dazzlingly entertaining out of sheer habit, and let's burn candles for Hylda Sims, who mixed beautiful poetry and skiffle. The following were as constant as could be, except in two or three cases where they weren't: Anthony Butler, Robert Wylie, Henry Moss, Jo Pye, Mike Pye, Lynne Leader, Gloria Dallas, Helen Leader, Janet Kerr, Reg Hall, Dave Arthur, Colin Harper, Brian Shuel, Sal Shuel, Simon Shuel, Mike Harding, Stan Kelly (ah, Stan…), Michele Coxon, Louise Eaton, Felix Eaton, Audrey Winter, Trevor Hyett, John Howarth, Martin Lynott, Ian A. Anderson, John Powles, Hans Fried, Belinda Hasted, Annie Hasted, Sean Davies, Gordon McCullough, Bob Pegg, Mairi MacArthur, Martin Carthy, Norma Waterson, George Knight, Karl Dallas, Carole Pegg, John Ellis, Joe Boyd, Mike Yates, Lea Nicholson, Peggy Seeger, Jeanie McLerie, Tom Leader, Annie Leader, Amy Harbour, Barry Seddon, Dolly Terfus, Phil Almond, Eddie Fenn, John Renbourn, John Cooper, Mel Purves, John Adams, Chris Coe, Pete Coe, Lorna Campbell, Leon Rosselson, Rosie Hardman, Marie-Claire Cadillac, Ríoghnach Connolly, Kirsty Almeida, Dave Swarbrick, Larry Kearns, John Alexander, Max Alexander, Andrew Cronshaw, Mike Walker, Tom Paley, Paul Peters, Marie Little, Marry Waterson, Tony Green, Shirley Collins, Stanley Accrington, Stan Ellison, Simon Prager, Graham Price, John Wilson, Donal Maguire, Bob Davenport, Thomas Stern, Simón Martínez, Adam Grant, Andy Kenna, Reinhard Zierke, mudcat.org, dubjax 1-32, Matthew Ord, Rod Stradling, Mick McElvaney, Sean McGhee, Kev Lee, Ruth Lee, Graham Jones, John

Fagg, Matthew Fagg, Alan Fearnley, Martin Kochaney, Michelle Leigh, Morag Rose, Timothy Fagan, Tim Stenhouse, James Douglas, Miriam Douglas, Pete Heywood. Above all thanks go to Bill Leader, too gracious to ever come on like a grumpy Johnson to a hapless Boswell, whatever the provocation.

FOREWORD

Keighley? Nobody knows where it is, or how to spell it, nor how to pronounce it, or so they say. I grew up here in the nineteen-fifties and sixties, so quite some time after Bill was in this area. But, in terms of access to live music, I don't think very much had changed between then and my time. A choice between traditional jazz, which was mostly competent but repetitive and boring – *I don't want to slag those musicians off, but my god it was boring* – or folk music, usually in dingy upstairs rooms in pubs. My friends and I gravitated towards the latter, partly because it seemed more authentic, and it was something you could share with the performer in the room. The performers were local and amateur rather than names to be conjured with. However, many were very good: beautiful voices, male and female, very pure and no microphones. Incidentally, with reference to the link between folk music and socialism, I became a Young Communist in the same upstairs room where our folk club took place. *I did! You had to be sworn in. It was a big deal. I was 16 years old. I became a Young Communist in the same room where we had our folk club! I mean, that's what they were there for. Pubs originally were meant for weddings and seventieth birthday parties and stuff like that. I think I've told you before, we'd meet on a Wednesday evening, if it was Wednesday. This one-armed guy would come over from Halifax and tell us what jumped-up middle-class burkes we were. He used to harangue us. I'd never been harangued. This one-armed madman from Halifax. That was in the Star Hotel on North Street.*

My teenage years in the nineteen-sixties were blessed by a wealth and variety of new music, much of it influenced by folk (and blues). However, an LP cost almost one fifth of my weekly wage: £5 in 1965, as a junior clerk in a textile office sitting each day on a high stool. London might have been swinging then, but Keighley wasn't. We bought our LPs in a white goods store called Ramsbottom's because it was the only source we had. Levi jeans were £3, and we bought them in an industrial clothing shop. No boutique here! If I bought a pair of Levi's, to last me two years, and an LP record in the same week, that left me with about £1. *And I'd probably give that to my mother for keep. That's the world in which we grew up, and that's the kind of austere world Bill was growing up in a decade before me.*

I don't play a musical instrument but that puts me among the majority – a listener. One thing in Mike Butler's account that particularly struck me was Bill's guiding principle to capture something of the actuality of being in the presence of the performer. I thank him for that, and I thank Mike for explaining it to me. *I know nothing about musical arrangements and recording, or music itself. I'm just a listener.*

Our access to live music may have been restricted, but we could get hold of more or less anything we wanted. There was always Bradford and even Leeds. Also there was a generous system of friends, lending records and books to each other. *(That's how I got to know Youngfellow, part of my education as a hippy.)* Perhaps the limitations added to the value of what we could get hold of.

Bill's recounting of his arrival in the town of Keighley, in the middle of the night, an evacuee aged ten *(or whatever he was, eleven)*, nervous, perhaps even a little afraid, moved me close to tears. So too Bill's memory of opening the curtains in Long Lee, to see a riot of green hillsides and trees. That was where I grew up. That was my playground.

I find your book fascinating. I'm going to read a bit more of it tonight.

Chris Ackroyd, dictated to Mike Butler, October 4, 2020

PREFACE

Everyman, I Will Go With Thee

Put simply, Bill Leader is Everyman. His story is the story of everyone buffeted by the large impersonal forces of history. No-one is immune, but as someone blessed with a long life, Bill has been buffeted by more than most, from the trauma of the Second World War to the collective calamity of 2020, the plague year, not to mention the imposition of VAT in 1973.

Wait a minute, I'm getting this wrong. VAT wasn't a disaster but a change for the better.

"VAT was much more easy to handle than Purchase Tax. If you didn't get registered (and very few people were, because the government wanted to limit the places they came and collected from), you had to pay Purchase Tax on something when you bought it. As a shopkeeper, for instance, you paid it when you bought it, and it stood, Purchase Tax-paid, on your shelf until you managed to sell the bloody thing, which could be forever. Whereas VAT was a much lower percentage and much easier to administer."

Now I notice it led to a reduction in the prices of LPs on Bill's Trailer and Leader labels.

"Yes. With Purchase Tax they decided how wicked you were to buy things like records. You were allowed to buy food. You didn't have to pay tax on food, and you didn't pay tax on kiddies' clothing, but adults' clothing, and all sorts of other things, had different rates of tax. It was a varying scale, whereas VAT was just an added sum for virtually every purchase that you made, irrespective of what it was, except for food and children's clothing. So they collected less from a broader range. It was less puritanical. 'That will teach you to buy bloody records'. (With Purchase Tax you were going to have to pay 70%.) So it was a democratisation, I suppose, of tax. It made things easier and it made things cheaper. If you were involved in this sort of thing, it made it easier for you to handle."

Right, but it's an immutable rule that misfortunes come in three, so I cast about for a third catastrophe. What about the vinyl shortage of the seventies?

"Well that was very brief. It was very inconvenient not to have new records pressed because someone in the Middle East was blowing up oil wells intended to make plastic for records, and drive cars, of course, but it was not quite the same."[1]

Bill promises to go away and think about it. Neither of us can bring ourselves to mention the dread 'B' word.

I started the book with the aim of gushing about rare records, but found myself writing a *de facto* social history, examining crosscurrents between public life and music, particularly folk music. I am now convinced that the music itself is a call to action. The Workers' Music Association sought to harness music to the struggle for social justice, and the WMA is where Bill got his start. All the scare stories in the right-wing press about the 'red' infiltration of folk music, which appear so quaint to modern sensibilities, flattered the cause by taking the threat seriously.

It was pianist John Ellis who told me that Bill Leader was working at Salford University. I was surprised. An exact count is complicated by misattributions, omitted credits and a flood of reissues and repackages, but at a rough estimate Bill has been responsible for engineering/producing some 500 original LPs of essential folk music. His output ranges from field recordings of traditional singers to charting the first steps of their successors in the folk revival. Without Bill's efforts there would be fewer songs to sing. He got to the Watersons, Bert Jansch, Anne Briggs, Nic Jones, Christy Moore, Gerry Rafferty, Martin Simpson, Billy Connolly and Ríoghnach Connolly before anyone, demonstrating that homemade intimacy (Bill's home, or sometimes their homes) was not just a means but an end. He was the ubiquitous recording engineer for Topic Records, the UK's oldest independent label, and freelanced for Transatlantic Records, a more commercial enterprise, before striking out and founding three labels of his own. Leader was dedicated to traditional artists and Trailer gave a platform

1 Bill Leader, interview with the author, July 29, 2020.

to the best club singers of the day. The punning names typify Bill's sense of humour (dry, subtle). The rationale behind a third label, Hill and Dale, was more pragmatic, but we'll come to that in its place.

The truth is, I didn't yet realise the impact John's news was to have on my life. At the time I stored it in that part of the brain reserved for trivia such as composer Philip Glass drove a taxicab, or Albert Einstein was a clerk in a patent's office. I associated him with a London address, and knew it by heart from the backs of Trailer sleeves – 5 North Villas, NW1 – and was vaguely aware of a later workplace, somewhere near Halifax, again from LP covers, but the vicinity was a shock. Bill seemed to creeping ever nearer. Without thinking about it, I suppose I presumed retirement in a Mediterranean resort, which is what I understood successful record producers did; or, more likely, I assumed he was dead. That happens too. But no, Bill Leader was alive and well and lectured in sound engineering at Salford University. He lived just down the road (albeit a long road) from John Ellis in Middleton, Greater Manchester. John also said Bill was a regular at the folk evening at the Oddfellows Arms, a pub in Middleton. I promised myself a visit. It had to be a superior session to attract Bill Leader.

And so matters stood in August 2009, when I received a press copy of *Three Score and Ten*, the official history of Topic Records. I pitched the story to my editor at *City Life* (the proud Manchester 'what's-on' was then a supplement of *Manchester Evening News*, with the resident jazz writer attached). This is what I wrote:

> Topic 70th Anniversary. The North West connection. Topic Records, the premier folk label, celebrate their 70th birthday this month. A sumptuous book/7 CD set is released to coincide with the anniversary. I have a copy. The label's pioneering engineer/producer Bill Leader – who recorded those early Bert Jansch and John Renbourn LPs in his living-room – only recently retired as head of Audio Dept at Salford University. An interview? His name appears on more classic albums than Phil Spector and George Martin combined.

("There is no audio department at Salford, and I was never the head of anything," Bill tells me now, but that's how I pitched it.)[2]

2 Bill Leader, amendment, November 4, 2019.

My editor was willing, but his reply struck an ominous note: "Happy to take 800-word interview with Mr Leader re. Topic. If he had a few vintage pics of him with artistes that would be great. Would pay you £70 but make haste, I only have 23 working days left here!"

Shortly afterwards *Manchester Evening News* moved from its opulent offices off Deansgate to an industrial estate near Oldham.If it has splashed any newsprint on jazz or folk since, it hasn't been under my byline. But I digress.

The point is, 800 words proved quite inadequate to the scale of the task. This was when I realised a book would be required. And what a book! On consideration, 500 LPs is an underestimate. Stan Kelly, one of Bill's comrades, when I asked how many children he had, evaded the question by discussing *Finnegans Wake*.

"The main character in *Finnegans Wake* is called HCE, 'Haveth Childers Everywhere', or 'Here Comes Everybody'. You haven't read *Finnegans Wake*? Nobody has actually."

Then, addressing the enquiry:

"Reports are still coming in".[3]

The same might be said about Bill's discography (it gained a new entry last week). His first studio, so distant he has difficulty remembering it, surely dates to his days at Films of Poland. His chief recording arena in the early days was the living-room of his modest flat in Camden Town, breaking the convention that if you wanted to make a record, you went into a studio. Transatlantic chief Nat Joseph honoured this rule, and so did Joe Boyd, who, asked to pick between two mentors, Rothschild and Leader, gives Paul Rothschild pole position for this reason:

"I had this very Rothschild-influenced view that if you had the money and you could afford it, you'd go into a studio."[4]

Bill followed the sound of a different drum.

"My idea was you set up the microphone in the place it was happening, if possible."[5]

So he went off to record the grand airs of Connemara *in Connemara*. After recording Paddy Tunney in his Camden Town living room, the traditional singer returned the compliment by offering the resources

3 Stan Kelly, interview with the author, November 20, 2012.

4 Joe Boyd, interview with the author, February 18, 2014.

5 Bill Leader, interview with the author, October 11, 2019.

of his (Paddy's) living-room in Letterkenny, Donegal. He documented Scottish travellers in Scotland, and bundled Jeannie Robertson into a clothes cupboard in Aberdeen. Pipers piped in between bouts of haymaking in Northumbria and reels and jigs played in strict tempo expressly for dancers were recorded in a room in a pub in Fulham, reinforcing the significant relationship between London pubs and Irish music. Leader (the label) issued historic recordings (by Stephen Baldwin and Cecilia Costello) and new recordings by old-timers like nonagenarian Dorset singer Charlie Wills and first-generation industrial folk-singer George Dunn. One Leader album, *Unto Brigg Fair*, collected field recordings made by Percy Grainger in Lincolnshire in 1908, and the mighty Joseph Taylor became a commercial proposition for the second time in a century!

Leader Records were simple in substance, lavish in presentation and timeless in nature. Trailer, modest in design (usually) but still handsome, pinpointed the moment of artistic (if not commercial) breakthrough of conspicuous talent; some would transcend the boundaries of the clubs and some had no wish to, but Bill invariably obtained a spark that eluded slicker productions.

I started this venture with little more than a layman's knowledge of folk and ignorant about the technical side of making records, but I reckoned it would be fun to have Bill as guide. My fumbling toward knowledge is described in the Introduction, where I accost total strangers to ask if they know Bill Leader (they don't) and chase someone pointed out to me as the second Mrs Leader across a crowded auditorium (and lose her).

Bill documented folk music when it was neither fashionable nor profitable. It still isn't. Here in England we're taught to despise our own ethnic music. It can feel like official policy when a Culture Minister openly sneers at folk music. Kim Howells (Labour), you'll remember, declared that his idea of hell was three folk singers from Somerset in a pub; and that awful, awful line about trying everything once except incest and morris dancing is repeated *ad nauseam*, and still people laugh. The gag is attributed to Sir Thomas Beecham, a convenient establishment patsy. Such are the insidious methods of cultural disinheritance.

Bill's time with the Workers' Music Association (with its affiliated label, Topic) coincided with the adoption of folk music by the radical left as *the* medium with which to celebrate the workers' struggle. Bill had a head start. His father, Bill Snr was a factory worker, a trade unionist and agitator who did what he could to make the world a better place. His mother, Louisa (always called Lou), a factory worker during the Second World War, would arrive early on the morning shift to sell the *Daily Worker* to the departing night shift.

So Bill was always aware how heavily the dice were loaded. That was something I had to find out: that I'm still finding out.

Along with all the ancestors and descendants that seemed to fit, the reader will find Thomas and Patrick… The reader will find a whole slew of Thomases and Patricks, in fact, but specifically the Irish brothers who came to England in the 1850s, as escapees from famine. Between them they founded the English line of Leaders, which has since spread to the Antipodes and is into its fourth generation in the USA. We shall mention Lou's brush with Judge Coleman of Baltimore – and, by two degrees of separation, with Saccho and Vanzetti – and bring ourselves up to date with Annie Leader, a singer in a punk band. Annie is Bill's youngest child. His eldest maintains the traditional family name – albeit Tom rather than Thomas – and followed his father's line in sound engineering. We have a family saga on our hands. That can bump up the word count too.

Then there are other, non-familial voices. The text is seamed with voices. *Sounding the Century* is part oral history, part masterclass, part-Pinter dialogue. Once you start talking to people, cracks in the orthodoxy begin to appear. Interviewees are extremely forthcoming with secrets and heresies; you only have to ask. Unfortunately, at such moments they will invariably say, "Of course, you can't print that." Still, my hope is that some countervailing opinions creep into these pages. If I have a wish, it would be to remove the stigma attached to radical socialists. As I hope to show, these are just the people – be they household names or anon – who made folk music what it is today, with Bill Leader foremost among them.

I lived in an attic in Madrid for a year during the writing of *Sounding the Century*, surviving on a diet of traditional song (*The Voice of the People* in permanent shuffle mode on my iPod) and, for entertainment, old British black and white films on the YouTube channel Dubjax. Some, like *The Common Touch*, from 1941, turned out to be directly relevant to Bill's story. Not much culturalisation going on there, you say. To which I reply, there's nothing like a tragic ballad by Joseph Taylor to underline the universal nature of *duende*.

The enforced isolation of the Coronavirus pandemic was a further fillip to the writing, and my concern for cultural survival shifted to concern for Bill's personal survival (my subject has turned ninety). I told him so, and this is what he said:

"We did go the Lowry a couple of times last week, but we didn't get very close to anybody, and the only person coughing in the auditorium that I could hear was me. It doesn't get to me because I'm going to die anyway fairly soon. I'd rather not be short of breath in quite the same way… It does sound a bit unpleasant, but then, you know, if it wasn't that it would probably be cancer in a year or two. Others can worry as much as they like but there's no reason for me to get uptight about it.

"*I'm quite looking forward to it.*"[6]

6 Bill Leader, interview with the author, March 16, 2020.

“Bill! How long has he been there?”
“From the beginning I suppose.”

Ivor Cutler, ‘The Dirty Dinner’

INTRODUCTION

Gold Folk

"I'M TRYING TO FIGURE out what to do about this gold medal thing," says Bill, referring to the highest honour in the conferral of the English Folk Dance and Song Society – a Gold Badge. "Apparently it's up to me if I want to put any event on. It's usually awarded to people who have some performance skills. Things like the Nic Jones concert a few weeks ago…" (The present tense of this conversation is October 2012.) "They said, 'It's up to you. We can send it through the post in a plain envelope so that nobody need know, or you can have a bit of a do, and we'd quite like you to make it available for EFDSS members to attend.' I think the implication was that Cecil Sharp House would be available free of all charge.

"My first thinking was we'll take Cecil Sharp House and do an Oddfellows evening down there and shock the bloody lot of them." The Oddfellows Arms is Bill's local in Middleton, and they have a session there every Monday night. "This is the living tradition! This is what's happening regularly and frequently! But then I thought of the logistics of getting everybody down there, and WAGs,[1] and accommodation, and transport, and getting them all back, just for the sake of shocking a few EFDSS members who wouldn't turn up anyway. It's not a good idea. So I thought of something local. I still like the idea of basing it around Oddfellows, on the basis that anybody who has walked in and done a turn there could be invited

1 The acronym – it stands for 'Wives and Girlfriends' – is no lapse into sexist terminology but the official, self-designated term. A sign to that effect adorns three joined-up tables facing the long table (ditto) where the musicians sit in a circle. It's not that the performers are male and the WAGs female, although that tends to be the general pattern. I've sat there myself when the Oddfellows has been full, and I've always received a friendly reception.

to sit among the regulars, and we'll try and make an evening of it."[2]

The recognition is long overdue, you might think, for someone who has been making great records for so long. Bill spans the medium from Charlie Wills to the Woodbine and Ivy Band and has literally hundreds of credits to his name. No honouree is more worthy. Performance skills are good, for sure, but backroom skills are just as necessary.

"I've got a year to do it, which means we can put it off forever. I don't want to do that. I'm trying to replicate a version of an Oddfellows night. I'll cheat a bit to make sure Ríoghnach [Connolly] is there, and Kirsty [Almeida] is there. I'll ask Stanley Accrington. So it should be quite a fantastic night. I thought Mike Harding might do the compering.

"My next step is to talk to Cecil Sharp House to see if they have any requirements, because they're presenting the medal. Shirley Collins presented it to Nic, but that was happening at the headquarters. She's the President of the Society. It used to be Princess Margaret. Now it's Shirley Collins. That's a huge indication of the way it's changed. The Vice President is Eliza Carthy. It would never have happened in my day, I can tell you. Douglas Kennedy must be spinning in his grave."[3]

From its beginnings as the Folk-Song Society, founded by Cecil Sharp in 1898, through to its amalgamation with the English Folk Dance Society (founded 1911) to become the English Folk Dance and Song Society in 1932, the avowed aim of the EFDSS has been to celebrate indigenous song and dance. Douglas Kennedy took over in 1924, upon the death of Cecil Sharp, and was director until he retired in 1961.

In fact, it was an old family business, "established 1898". Douglas' mother, Margaret Kennedy-Fraser, set the tone by turning crofters' songs into art songs fit for the Edwardian parlour with such tomes as *Songs From the Hebrides* (three volumes, 1909-1921), *From the Hebrides: Further Gleanings of Tale and Song* (1925) and *More Songs of the Hebrides* (1929). Maud Karpeles, Cecil Sharp's formidable assistant on the 1916 song collecting trip to the Appalachian Mountains and a fierce protector of his legacy, was Douglas' wife Helen's sister. Peter was Douglas and Helen's son (hence Maud's nephew), and became, by turn, a musician, broadcaster and song hunter. Because celebration and preservation

2 Bill Leader, interview with the author, October 5, 2012.

3 Bill Leader, interview with the author, October 19, 2012.

were part of the remit, the EFDSS was a very conservative institution, as opposed to the radical Workers' Music Association, which came along a bit later. Bill was always more of a WMA man. (I wonder if the tardiness of the award is explained by Bill belonging to the wrong faction.)

"So, I'm going to talk to them, and see what their requirements are, what stipulations they might bring to bear, and then slowly build it up."

A date, September 28, 2013, is fixed for the Gold Badge ceremony/ entertainment, now officially known as the Gold Folk Show, and a venue has been chosen – The Met, in Bury, Greater Manchester. I'm due to interview Stan Kelly for the book, and Bill asks me to invite him as a guest performer.

"I don't quite know if he's got the voice for a 200-seater hall, but I think he probably has, if he rises to it. But he will rise to it. He's a man who rises."[4]

(Concertina player and Morris dancer William Kimber got his Gold Badge in 1929, but he was a friend of Cecil Sharp. Bill was busy being conceived in 1929, and, at the close of the year, busy being born, and of course Gold Badges can only be retrospective awards for longevity and achievement, and never advance forecasts. Yet it strikes that 2012 is still a little late in the day. Bill isn't worried. He only wants to make the Gold Folk Show as good as it can be. I'm consoled by the thought that Cary Grant received his one and only Oscar when he was 81.)

And then, on the very eve of The Gold Folk Show, disaster strikes, and Bill comes down with a vile case of stomach cramps. Lynne rushes him to the Accident & Emergency Dept. It is one a.m.

"I'm probably not going to make it on Saturday," Bill tells Lynne.

"No, I don't think you are," returns his wife, sorrowfully but firmly.[5]

There is still so much left to do. Bill, in discomfort and with frequent breaks for the loo, scribbles down the titles of songs, the names of singers, the programme order, and a stage seating plan, with pencil and paper provided by a nurse. He is clear about the combinations of musicians he wants, and the combinations he doesn't want.

"Trevor would be really good at doing this. Can you make sure he does this bit?" (Trevor Hyett honed his skills on-set in the studios of Granada TV.)

4 Bill Leader, interview with the author, November 19, 2012.

5 Lynne Leader, interview with the author, June 24, 2019.

"Amy could help you with that bit." (Bill's daughter Amy worked for the BBC during Greg Dyke's tenure as Director-General and is presently managing Hebden Bridge's community town hall.)

Bill's most controversial decision is to banish microphones and rely on a natural acoustic. On the face of it, this seems a denial of his calling: Bill's fame as a sound engineer rests on the strategic positioning of microphones. It's a big risk, and rests on the assumption that pub intimacy can be transplanted to a theatre space. The difference is one of scale. The bar is a very long way off, which is hard on pot man Henry Moss, fulfilling his traditional role of collecting empties and replenishing the performers' drinks. The stage has been lowered, but he still needs to exit, wander up the aisle and be ushered to the bar next door, then repeat the journey with tankards of flowing Guinness. Landlord Darren Potts and barmaid Debbie smile and say hello, which is not half of it normally, and the Monday-night crowd, asked only to be themselves, look small and helpless and unprotected as they take their positions.

Bunuel tapped into this wellspring of irrational fear in *The Discreet Charm of the Bourgeoise* when diners, ostensibly in a restaurant and about to start an endlessly deferred meal, suddenly have the curtains yanked aside and find themselves on stage with an expectant audience sitting in front of them. In short, it feels odd to be an 'Oddie'. I'm grateful I'm only an extra.

Mike Harding notices the absence of sound transmitters.

"I haven't got a mic. We need mics. Where's the soundman? I'll get it sorted!"

Lynne says that Bill has requested a natural acoustic. Mike takes it in his stride. Whatever Bill says is alright with him.

Two honorary 'Oddies' are on the next table. Stan Kelly is having a senior moment and can't remember the composer of *Hugh the Drover*. The lapse really seems to vex him. Dave Arthur racks his brain and takes a stab at Benjamin Britten, before the two none too confidently settle on Alan Bush. Arthur, Bert Lloyd's biographer, remembers interviewing Alan Bush.

"I saw Alan Bush when he was in his nineties. He was going blind but still working. I went up to see him in his house in North London, and he was crouched over this enormous table, with the manuscript of his last piece, and he had this huge magnifying glass. It was unbelievable. The persistence. Unfortunately, he couldn't remember Bert. His memory had gone. I said, what did you do with Bert when you were working together? He said, did I know him? I said, yeah, you knew him… He said, oh what else did I do? I told him. He said, did I? It was one of those bizarre reverse interviews where I was telling him what he'd done, and he was going, oh really?"

Dave Arthur is here as the official representative of the EFDSS, and Stan, an admirer of *Bert* the book, is pleased to meet him, although he insists on calling him 'Arthur' or sometimes 'Douglas'.

"For the longest time you were both Douglas Arthur and Arthur Douglas," he says ingratiatingly, "and I couldn't remember if there was a comma missing. Did you have a choice in the matter?"

Dave, undaunted, bounces back:

"Arthur Douglas sounds very Wildean. You're confusing me with Oscar Wilde."

"I can't remember the meaning of amnesia," says Stan extricating himself from a tight spot with charm, as he is wont to do. His body may be crumbling, but his wit never deserts him. The talk moves on to Bert Lloyd's great rival, Ewan MacColl.

"I had a great shock when I went to Ewan MacColl's house," confides Stan. "Before he married…"

"Peggy."

"Littlewood."

"Joan Littlewood, yeah."

"His first wife. All that Habitat furnishing. He was a rouble millionaire at the time. From his Russian escapades."

"Really?"

"His songs were selling in Russia."

"I didn't know that."

"And at times he had to go over to Russia to spend the money."

"Yes, you couldn't bring it out in those days, you had to spend it there."

"I've always had mixed feelings… He wore his heart on the left and his wallet on the right. Do you know that? It comes over better in French."

Stan turns from the felicities of the French language to the perplexities of Cherokee as the audience files in.

"It's hard to distinguish nouns from phrases. 'Cemetery' is literally 'you go there and you don't come back'. That's agglutination."

"I'll just go and tune my guitar," says Dave, and absents himself.

I can see John Renbourn and Jacqui McShee in the fourth row from my vantage on stage. It seems a reversal of the natural order somehow.

Stan turns to Derek Schofield, who has joined the table. Schofield, like Arthur, is a former editor of *English Dance and Song*, the EFDSS house journal.

"Do you follow Nic Jones' story?" Stan asks.

"Yes, yes," says Schofield.

Michael Proudfoot's documentary, *The Enigma of Nic Jones*, with a contribution from Bill, had aired the previous evening on BBC Four. It had impressed Stan.

"Bill gave quite a long interview. Because apparently, he was associated in the very early days. The Halliard. Nic Jones was the leader or founder or co-star of The Halliard. He's always been a hero of mine although I've never quite met him. It was quite a pleasant shock to see Bill, looking very young actually."[6] (Stan is Bill's senior by four months.)

The music strikes up. Twin fiddlers Martin Lynott and Steve Keene lead the opening medley of jigs, and Martin Hall moves things along with his banjo. These three are the backbone of the Oddfellows rhythm team, whilst guitarist Ian Reynolds picks his own route through reels, jigs and breakdowns. True to the Oddfellows spirit, everyone continues chatting. Independent of the music, a lone person begins to clap. Someone else takes it up and the ripple swells to a rousing ovation. I look and see the cause: Nic Jones is taking his seat in the fourth row. He flourishes a walking stick by way of acknowledgement.

Karl and Gloria Dallas make their entrance unnoticed by all but me.[7] Some applause might be appropriate, because Karl has been actively producing journalism, books, songs and plays for half a century and more; his screenplay about the assassination of Russian journalist Anna Politkovskaya is in the slush pile of a famous Hollywood producer; one of his songs, 'The Family of Man', is a certifiable folk standard. His

---

6 Stan Kelly's table talk with Dave Arthur and Derek Schofield, The Gold Folk Show, Bury, Manchester, September 28, 2013.

7 Not quite. "The family section of audience all noticed their entrance – it was clear that Gloria was less mobile than we realised, so I nipped down and found them a seat at the front, so that she didn't have to climb stairs to the seats we'd allocated them" – Lynne Leader, annotation, May 30, 2020.

energy is prodigious. Yet he can be overbearing with it, so any cheers might be mixed with catcalls. Gloria is beside him, as ever. She is too frail now to fight against her fate of being overshadowed by the men in her life. This was the cause of friction in her marriage to Bill (Gloria is Bill's first wife), but she is devoted to Karl. Gloria turned on Prime Minister Tony Blair in a TV debate about Iraq during Karl's time as a human shield in Iraq. Why, she demanded to know, was her husband in greater danger from the British than from Iraqis?[8] Now frail and stricken by Parkinson's Disease, Gloria remains indomitable.

The show must go on, even if the principal is missing. John Howarth, unofficial master of ceremonies, is reciting Robb Wilton's 'Back Answers', a monologue in rhyme about life's unpredictable twists. Some of these are benign – like mistaken marriage to a wealthy woman (something about identical twins; the pay-off is "never mind, she'll do!") – and some are calamitous, as when the captain of a sinking ship gets protective about places on the lifeboat:

> *I know it's upsetting, but what's the use fretting?*
> *We might have lost all of the crew*
> *But now, as I say, we can all get away*
> *And we'll only lose one... And that's you!*

I imagine Bill as a vernacular Professor Xavier, guiding the proceedings by telepathy from a ward in North Manchester General Hospital.

Stanley Accrington is charging up and down the stage in a helmet with cow's horns, waving a wooden sword and declaiming in Old English. Stanley entertains Oddfellows patrons with an original topical song every Monday. If collected, these would constitute a sharp commentary on our times.[9] I have never had him pegged as an Anglo-Saxon chronicler.

---

8 https://www.youtube.com/watch?v=hzAOwfj8lII. A real Diana Gould moment, this.

9 Lo and behold! "What I have done is condense my output of 15 cassettes (1981-1997) and 8 CD's (2000-2018), a total of 236 different tracks, into a single 'sound DVD' which contains all the tracks as mp3's that can be downloaded onto a computer and played on whatever music player it has. The price for this collection is £20, which makes it less than a pound per album, or under 10p per song. If you would like to invest in this sound bargain, let me know. Cheers, Stanley A." Email from Stanley Accrington, November 27, 2019.

'The Battle of Maldon' commemorates a defeat at the hands of Vikings in 991 AD. Says Stan: "In 1992 I was booked to play at Maldon Folk Club, and, doing a bit of research, which I always do when I go somewhere I've never been before, I found out all about this battle. They'd just had the thousandth anniversary of it the year before. It coincided with a friend of mine, a very eccentric character, sadly dead now, from Halifax way, who used to do the Albert and the Lion monologue in the style of Beowulf, which I thought was wonderful. I thought I'd love to write something in Anglo-Saxon, as it were. So I wrote this long poem about the battle, with theatricals and all sorts of stuff, as a one-off for this club. It happened to be the club's last ever night, which was another notch in my guitar. I kept it in the act. It became one of my party pieces. Then, about ten years ago I got an email from the Essex County Archive at Chelmsford, and they said, 'We understand that you have a version of the poem about the Battle of Maldon.' They said, 'Can we have it? We can't pay you any money, but we'd like to keep it in our archive.' I said, 'Why would you want it?' They said, 'Because we've got the original 11th century manuscript.' (There is an Anglo-Saxon poem about the Battle of Maldon.) As I say when I introduce it, scholars of the 28th century when they go into a county archive (they won't go to a county archive: they'll call it up on their computers or whatever), and say 'Can I see the Battle of Maldon poem?', they'll say, 'Which one do you want to have a look at?'"[10]

With the battle over, Andy Kenna strikes up his squeezebox and announces something in his very distinctive Liverpool-Irish brogue. (I could scarcely believe it when someone told me he taught French.) "Normally I'm the token Scouser," he declares casually, "but I'm pleased to say that tonight we've got another one here."

In the first serious departure from Bill's schedule, Stan Kelly mistakes this as his cue, and instantly starts to sing…

Oh, you a mucky kid, dirty as a dustbin lid…

Kenna silences his squeezebox immediately.

The mucky kid featured in Stan's discourse to Dave. Dave was saying how Bert's son, Joe, a heroin addict, died of an overdose in Better Books, a Charing Cross Road bookshop. Stan recalled that he had come across

10 Stanley Accrington, interview with the author, June 10, 2020.

one of his own books in the second-hand section of Foyle's in Charing Cross Road. It was inscribed to his son David as a birthday present.

"I never told him. I just bought it. Isn't that sad? Oh, weep, weep. I hope he spent the money wisely."

"He must have been broke. He must have been desperate, I'm sure."

"So it's my fault?"

"If only you'd given him a bit more money?"

"But this David is the original 'mucky kid', my second son."

Despite Stan Kelly's frailty, his voice rings out, unaided by amplification. 'Liverpool Lullaby' is raw in its maker's own account and has none of Judy Collins' purity or Cilla Black's sentimentality. It gains from his closeness to 'the mucky kid'. It doesn't matter that Stan gets the words wrong. He always gets the words wrong. Possibly he got the words wrong when he wrote it. That is his prerogative as creator.

When the applause dies down, Andy Kenna picks up where he left off. 'Has Anyone Seen Grandad?' is an Alzheimer tragicomedy by Bob Watson. A jig from the ensemble follows, and everyone starts talking again. Another Oddie regular, Ian Sidebotham, tackles Ian Campbell's 'The Old Man's Song'. This makes three parent-sibling songs in a row by my reckoning (although 'Liverpool Liverpool' did appear out of turn). 'The Old Man's Song' is written from the perspective of an old soldier who has outgrown every purpose except relaying the lessons of his life, but no-one is interested. The song nails the cyclical pattern of history. Sidebotham sadly notes its relevance today.

To pick up the tempo: Desi Friel essays 'The Rising of the Moon', an Irish rebel song. Stan joins in lustily on the chorus. Kirk McElhiney traces the romantic fatalism of Bert Jansch's 'Tell Me What is True Love'. Romance gives way to lust when James and Jay Gansler offer a duet, 'In Spite of Ourselves', a John Prine song made for dirty laughs. Jay, incidentally, has the world's greatest dirty laugh, but it's kept under wraps tonight. Ríoghnach Connolly, a singer from Armagh who has made her home in Manchester, offers a moving 'Sweet Inniscarra'. John Howarth reprises 'A Mon Like Thee', a relic of Lancashire music hall, part shaggy-dog and part wish-fulfilment. Ian Reynolds leads a lively 'Salty Dog' and closes the first half with 'Will the Circle be Unbroken'.

It's the interval, and I've been ordered to – dread word – 'mingle'. Here, assembled in the same place for one night only are people from different phases of Bill's life, though naturally they thin out the farther back you go. Audrey Winter, widow of folk journalist Eric Winter, remembers Bill arriving in London from Bradford in 1955.

This presents a would-be biographer with a heaven-sent opportunity. I have a bespoke business card at the ready. Along with my contact details, it reads, in forty-eight-point uppercase italics – *MIKE BUTLER, BILL LEADER BIOGRAPHER* – and, in eight-point condensed italics, this inspirational quote:

There are many less worthwhile things one could be doing.

This is what Andrew Cronshaw told me, when I rang to confirm the existence of the Trailer EP *Cloud Valley* (LREP1). His instinctive response, when I said I was writing a book about Bill Leader, was actually a wordless huzzah, instantly qualified by the remark quoted.

The first two people I approach, mainly because they happen to be in front of me, respond with a disappointing, "No, don't know him; never met him." I make a mental note to be more methodical in approach. The next person I buttonhole, a beautiful woman of indeterminate age, remembers seeing Bill in army uniform. This is more like it! I calculate the math and do a double take. If she knew Bill during his National Service in 1951-53, she can't be far off 80 herself. I miss the name. (Now I have it: she is Louise Eaton, the wife of Bill's friend and mentor, Alex Eaton.) The question of Bill's uniform will return to haunt both of us, by the way. The woman points out the back of Helen, Bill's second wife. I set off in pursuit but miss her in the crowd.

Carried by the flow now, I find myself face to face with Nic Jones. He cheerfully reveals that 'Canadee-i-o' fades because he fluffed the ending. This is interesting, but not to my purpose: 'Canadee-i-o' comes from the Tony Engle-produced *Penguin Eggs*. Later, upon investigation, I find that 'Canadee-i-o' ends firmly and doesn't fade at all. Be not too hard. Unreliable evidence is a common failing even in those who haven't had their motor and cognitive skills wiped out in a horrific car accident.

The tide sweeps me onward and I next bump into Julia Jones, Nic's wife and pillar of strength. She never liked folk music, she tells me, and is gone. On the stage, near my empty chair, I notice Stan Kelly and Karl Dallas are chatting. I would give a lot to know what the two old comrades and rivals-in-love are saying. Hunter Muskett play an unplugged set in the lobby and there are hillbilly musicians in the bar, but you can't be everywhere. An announcement requests we take our seats for the second half. I can do no more than nod to Audrey Winter, my first interviewee after Bill.

Audrey took priority by dint of seniority. We met outside The Barnsbury, a pub on Liverpool Road, London N1. I ventured inside just once, to buy a drink. The entrance to the cellar was on the forecourt, next to where we sat, and at one point her words were drowned by the sound of metallic beer barrels rolling through the hatch to the vaults. She warned me off. Bill, she said, is not the kind of person you write books about. She meant, I suppose, that Bill is not Ross Poldark, but the stipulation, it seems to me, sets an impossibly high bar for candidates for biographical treatment.

Now it's impossible to think unkindly of Audrey, and not just out of respect for age. She admits to running a "free hotel for the folksong world". Her first encounter with Barbara Dickson was typical...

"She had been singing at the Fairfield Hall in Croydon, and Eric had gone to listen to her, and he'd got home very late, and of course I didn't know she was there. Eric put down an inflatable mattress we had in the dining room. I was up first in the morning to go to work, and not knowing anyone was there, I walked in and fell straight on top of poor Barbara Dickson. I said, 'Who the hell are you?' 'Oh dear, I'm so sorry.' And then she laughed. It was alright. That must have been quite a shock for both of us. You never knew what was going to happen at my house."[11]

Audrey, a level-headed anti-Stalinist in the company of Uncle Joe worshippers, is wary of alpha males. She said the folk scene was "a small but interesting pond", and clearly feels that pond-life doesn't merit an official guide. Bill feels the same view. He was perturbed at being the centre of attraction when I first floated the idea of a book and agreed only when I promised to give due prominence to obscure contemporaries. He made a list of Possibly Significant People to aid my understanding. This was entirely compatible with my aim. I would like my book to illuminate the known, the unknown and the unknowable, and alert to some good music. It would be properly Marxist, coming from the bottom up. This is what I should have said to Audrey.

But now there are speeches and an award ceremony to be got through. Mike Harding takes centre-stage and begins by telling Bill's

11 Audrey Winter, interview with the author, September 6, 2012.

story in Bill's words, drawing on background notes supplied by Bill.[12] His comic timing and sly delivery blurs the boundary between what's funny and what's matter of fact.

"Aged 21-25: engaged in youthful political and musical activity in the Bradford area…"

This raises a mild titter, perhaps because of the subliminal connection between Bradford and Rochdale, as in 'The Rochdale Cowboy', Mike's big hit. A consummate comedian, Harding builds on this small encouragement.

"Missionary work!" he roars. The mild titter turns to unrestrained merriment, and there is even applause.

"…including starting a local branch of the Workers' Music Association with Alex Eaton, who went on to be one of the pioneers of the English folk movement. He also co-founded the Topic Folk Club, one of the longest running folk clubs in England. Aged 25, landed in London, which led to being involved in the production of Topic Records, serving an enthusiast's label, which has now become the longest established independent label in the world. Aged 38, progressed from in-house producer of Topic to freelance producer who decided to start his own independent labels with his wife, Helen. Aged 43, moved back north and started a recording studio in Greetland, near Halifax. Aged 43-53, maintained the Leader and Trailer labels and continued with valuable supplemental work with Transatlantic. Aged 53, contract with Transatlantic terminated by its new owners, leading to eventual loss of both Leader and Trailer labels.

"Aged 53-76, taught recording studio techniques at Salford College of Technology, and later Salford University, starting as a lowly lecturer and finishing as visiting professor."

Catching Harding's amusement at Bill's elevation, the crowd respond with a look-at-you "whoooo!"

"He wrote this, not me," chuckles Harding.

"Aged 76-83, after retirement coaxed back into the studio by Manchester musician John Ellis; awarded the EFDSS Gold Badge, enhancing possibilities of further recognition as a) being a visiting

12 "I sent him some bullet-points, thinking he would then elaborate using those as background, but what he did, the bastard, was he just read the bullet-points out." Bill Leader, interview with the author, February 25, 2016.

professor with absolutely no qualifications at all b) being the oldest functioning record producer or c) the deafest functioning record producer."

With material as good as this, Harding is unstoppable. The audience is in stitches.

"That bit is Bill. The rest of it is me."

And Harding, suddenly serious, advances his theory of the "ripple effect", highlighting Bill's role in the Irish folk awakening. Christy Moore is key in this. He was crashing on Mike's floor in Crumpsall when the men had a far-reaching conversation.

"I suggested to Christy to give Bill a ring. 'Do you think he would have me on the label?' he asked. I said I reckoned so. He did phone Bill, and Bill said he was more than happy to record Christy."[13]

The resulting album, *Prosperous*, was seminal. It single-handedly gave rise to Planxty. Without Planxty, there would be no Bothy Band, no De Dannan, no Danú...

Would there be Mike Harding without Bill? Undoubtedly. But Bill laid the groundwork for 'The Rochdale Cowboy' with *Deep Lancashire*, Mike's debut on record, and *A Lancashire Lad* (LER 2039), his first full-length album. The demand for *A Lancashire Lad*, on the small Trailer label, outstripped supply; a clear sign that Harding was headed for bigger things.

Dave Arthur reads a message from Shirley Collins, president of the EFDSS, who sends her apologies. Tom Leader, Bill's eldest son and a sound engineer himself, collects the Gold Badge on Bill's behalf.

We're back to the music. Dave Arthur sings 'Bill Dalton's Wife', originally a poem written by the radical Georgia poet and political activist Don West, father of ballad singer Hedy West. Much of what follows is familiar to an Oddies regular, although the rocking 'Hal-an-Tow' from James Gansler is new. The magic words 'Roll in My Sweet Baby's Arms' inspire the bluegrass contingent, headed by Ian Reynolds, to fresh heights. Stan Kelly amuses and puzzles with a parody of a Child ballad, 'Captain Wedderburn's Courtship'. It mixes riddles and washing powder.

The Gold Folk Show closes with two valedictions. 'They Don't Write 'Em Like That Anymore' treats of nostalgia for pre-rock 'n'

13 Mike Harding, The Gold Folk Show, Bury, Manchester, September 28, 2013.

roll songs. It was written by Pete Betts, a friend of Vin Garbutt from South Bank, a working-class neighbourhood in a working-class town, Middlesbrough. John Howarth has turned the song into an Oddies staple. If John will excuse the pride of a native 'Boro boy, *TDWELTA* belongs to Middlesbrough. Its secondary and tertiary concerns – booze and scatology – are the great 'Boro themes. 'Ee When I Were a Lad', another song in Howarth's repertoire, is its true North West counterpart. John's interpretation of *EWIWAL* is definitive.

'The Parting Glass' closes Oddfellows every Monday night, and its theme of companionship as the one consolation in our fleeting existence is especially appropriate tonight. Ah, but can we ever know our companions? Perhaps Karl Dallas is right to ponder Bill's impenetrability, and, by implication, the impenetrability of every one of us.

"He did a series on the radio,"[14] Karl tells me, "and it was all a great smokescreen to prevent anyone from finding out about the real Bill. I wouldn't say I knew the real Bill, although we drank together regularly."

"This is fascinating," I reply, resembling Tom in a *Tom and Jerry* cartoon, with the words *BILL LEADER BIOGRAPHER* branded on the outside of my protruding eyeballs in forty-eight-point uppercase italics. "Tell me about the real Bill."

"Well, I don't know!" says Karl, sadly. "That's the thing."

14 Karl Dallas, interview with the author, February 27, 2013. The "series on the radio" was *Leader's Tapes*, as produced and presented by Rab Noakes, an extensive interview aired over two editions of *Travelling Folk* for BBC Radio Scotland, and broadcast by Radio 2 in March and April, 1996.

"It's a quare world," Jamie said one night, as we sat in the glow of the peat fire… He took his short, black pipe out of his mouth, spat into the burning sods, and added: "I wondther if it's as quare t' everybody…"

ALEXANDER IRVINE, *MY LADY OF THE CHIMNEY CORNER*

I

Gas Works, Palaces and Earthly Damps

THOMAS AND PATRICK LEADER, brothers from Ireland, came to London in the 1850s. Their father, also Thomas, is the ultimate patriarch and Bill Leader's great grandfather on his father's side, and his great great grandfather on his mother's side. The salient fact is that Bill's father and mother were cousins. Bill's mother, Lou, had the same surname before and after her marriage.

"I don't think it's ever been illegal for cousins to get married," says Bill, in answer to my query. "It's not recommended in some quarters. They're about fourth cousins or something. They're pretty distant. The common forebear is three generations away in the case of my mother, and two generations away in the case of my father. So I think it's allowed. Whether it's advisable, time will tell."[1]

1 Bill Leader, interview with the author, April 10, 2013.

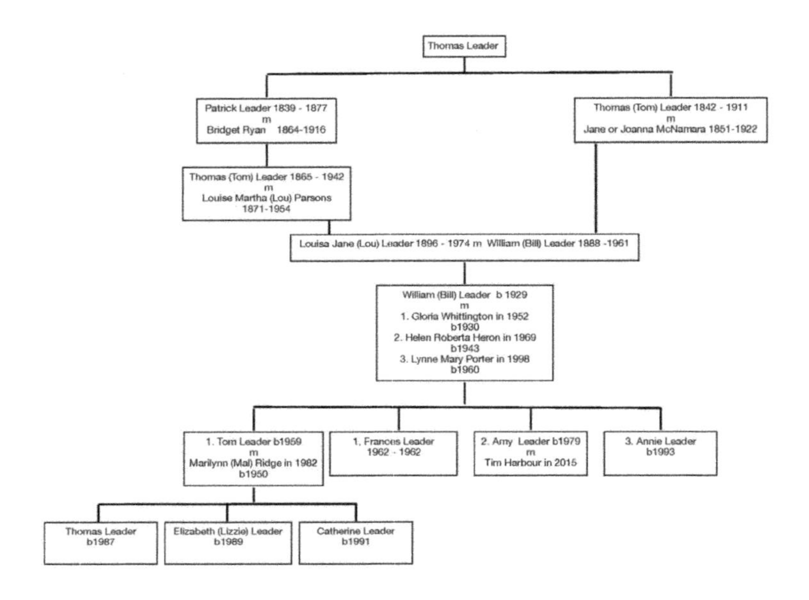

Fig. 1 The Leader Bloodline

The family came from Ennis, the county town of County Clare. That's what Lou said, and it is what Bill accepts. Others claimed Limerick as the family home, but no-one is quite sure whether they meant Limerick city or Limerick county. Sally, Bill's aunt, who died aged 91 "pickled in Guinness", went to Ireland to research the family tree and proposed a third candidate: Kilkee, on the coast of Limerick. She found a relative in Kilkee who was a Catholic priest. The matter was promptly hushed up in the proudly atheist family, and never mentioned again.

This Thomas and Patrick are the Romulus and Remus of the Leaders. Thomas married Jane (sometimes known as Joanna) McNamarra in 1873 and parented nine children. The sixth was Bill's father, who shall henceforth be known as Bill Snr, though nobody called him that in life. It is reluctantly adopted here to distinguish him from his son, the hero of these pages. Patrick married Bridget Ryan and sired three children, Thomas, Patrick and Bridget. The eldest, Thomas, married Louisa Martha Parsons on Christmas Day in 1889. They had eleven children, of which Lou was the fourth-born, making Thomas and Louisa Martha Bill's maternal grandparents.

Bill puts it in a nutshell: "My mother was at the elder end of her bunch of siblings, and my father was at the younger end of his bunch of siblings. Basically, there was a generation between them, though only eight years difference in age."[2]

The two brothers and founders of the line worked as stokers. The 1861 census finds Patrick Leader in Vauxhall Square, Lambeth, living with a family. His occupation is given as a 'labourer in a gas factory'. By the 1871 census he is living in Laundry Road, Battersea, with his wife and three children. The Battersea locale and occupation would suggest that Patrick was an employee of the London Gas Company, a small gas works in the area. The same census (1871) has brother Thomas' address as 10 Vauxhall Square, Lambeth (Patrick, in 1861, had been at no. 14). He is 26 and his occupation is given as a stoker in the gas works.

Stokers had the arduous and dangerous job of feeding the furnace.

> It was dhreadful hard work, and as hot, aye, as if you were in the inside of an oven. I don't know how I ever stood it. Be me soul, I don't know how anybody stands it… Standing there before them retorts with a long, heavy rake, pullin' out the red-hot coke for the bare life, the rake red-hot in your hands, and the hissin' and the bubblin' of the wather, and the smoke and the smell – it's fit to melt a man like a rowl of fresh butther. I wasn't a bit too fond of it, at any rate, for it 'ud kill a horse.[3]

When the London Gas Company was acquired by the Gas Light and Coke Company, Patrick was ordered to Beckton Gas Works, which had commenced operations in 1868. Beckton supplied the metropolis north of the Thames and was the world's largest gas works. A subsidiary, Beckton Products, marketed the by-products of gas manufacture, including coal-tar, ammonia, sulphuric acid and nitrate of soda. The furnaces and ovens were in constant use and steam engines hammered incessantly. Thick, pitch-black tar coated every surface. The geysers of steam that erupted underfoot were entirely man-made and beyond the realm of nature. To some this made Beckton a province of hell, while others saw it as the epitome of human progress. Even the pollution was miraculous. One worker, Mr Orchard, was so entranced

2 Bill Leader, interview with the author, March 2, 2012.

3 Henry Mayhew, *London Labour and the London Poor, Vol. II*, p. 87.

by the white crystals that formed on dried scoopfuls of yellow mud that he ran to show them to Mr Harris, the manager. And rhubarb grew in abundance. There was so much of the stuff that Beckton was affectionately known as 'Rhubarb Junction' among the workers. This, according to the memoir of A.E. Bigg, a Beckton employee:

> My mother was very fond of filling my bag up with many nice eatables, and when the boys were hungry I used to pull some rhubarb and make them some stewed rhubarb, as it grew all over the works, so that the place was called Rhubarb Junction, and anyone arriving at Beckton Station was asked the question, "Beckton or Rhubarb Junction?"[4]

Bigg's sanguine disposition failed him only once, and that was when dead bodies started to pile up on the foreshore. Beckton Gas Works lay on the banks of the Thames, and on September 3, 1878, the light collier *Bywell Castle* collided with *SS Princess Alice*, a pleasure boat returning to London Bridge after an outing to Gravesend. She sank with the loss of between 600 and 700 lives. Sheds at Beckton were turned into makeshift mortuaries. Bigg remembers: "There were sheds full of bodies, and many others were lying uncovered waiting identification, which was an awful experience for anyone. Some were never owned and some of the saved were rendered insane by the shock. Some children were saved but never claimed, and a few were adopted by some kind friend. Among the latter was Mr G.C. Trewby, manager of the Gas Works."[5]

Tar and pollution didn't deter the navvies from their raffish ways, it seems. Again, according to A.E. Bigg:

> Gangs of navvies in picturesque dress were excavating on the Tar Wells, and also Liquor Well; and it was so strange to see the navvies arrive at work with their coloured knee breeches and stockings, and sleeve waistcoats with long velvet backs to them, pipes sticking in their slouch hats. It took months to get these wells out.

Equipped with only native strength and the basic tools of shovel and pickaxe, navvies shifted earth on a seismic scale and dispersed it using wheelbarrows. The navvies, predominantly itinerant Irish labourers, built the canals and the railways and the London Underground system.

4 A.E. Bigg, 'Memories of Fifty Years', *Co-partners Magazine*, 1930.

5 A.E. Bigg, ibid.

They worked at piece-rate and were encouraged to destroy their health for that extra kilometre. Elizabeth ('Mrs') Gaskell in *Cranford* (pub. 1853) reveals how they were the fear of their host communities and describes the impact of the coming of the railway. The genteel ladies of Cranford (based on Knutsford in Cheshire) could take no comfort from the fact that the navvies' aggressiveness was entirely self-directed. Altercations, or situations where one man would seek advantage over his fellows, were usually settled by a blow to the head.

Their flamboyancy of dress comes as a surprise. Do knee breeches and tar wells go together? This is the question I put to Reg Hall, the author of *A Few Tunes of Good Music*. Navvies, declares Reg, had a lifestyle of their own.

"As you probably know they did the Siberian Railway, and they did all the South American railways. The same people. The last big bits that they did were the Underground cuttings and things. So those gangs of navvies were going from the 1800s, with the canals, right up to 1913-14. They had a lifestyle. And part of the lifestyle was dress. What else did they do? They drank. They ate a lot of beef. They didn't have women with them. They certainly didn't play the fiddle. They might have sung traditional songs. Who knows?"[6]

A lot of them had names like Tom and Pat.

Jo Pye, Bill's cousin and the *de facto* Leader family historian, declares, "There's a hell of a lot of Thomases in our family, and a hell of a lot of Patricks."[7]

Jo knows more about these Toms and Pats than any person alive.

"There were just so many of them," she says. "It's very confusing. My father [Lou's brother] was Pat but was generally known as Ginger Pat because he was the only one that was. And Uncle Bill's brother was called Uncle Pat, and then there was Cousin Pat who lived around the corner somewhere. His father was grandfather's brother Pat, and then there was Young Pat, who happens to be a girl. When we were children, we called everybody auntie, and then when we got older, we called them cousin, and then they just became Sally and what have you."

"You'll need a genealogy expert to do the Leader family. It's impossible," grumbles Mike, Jo's husband, the former Lord Mayor of Sheffield; Jo is the former Lady Mayoress of Sheffield (2004-2005).

6 Reg Hall, interview with the author, April 18, 2016, for this and subsequent quotes.

7 Jo Pye, interview with the author, December 12, 2013.

"Yours is more complicated," retorts Jo. "You just choose not to know about it."

Jo's tone towards Mike might be described as waspish, but there is love behind it. Mike has a point. The limited stock of names in circulation in large Anglo-Irish working-class families of the period can be a headache for the genealogist.

"It's not as bad as my mum's family," says Jo, whose research extends to her maternal side, the Wicks.

"She had an Aunt Frances who married a man called Francis. They had four children, two of whom were called Frances and Francis."

The musical tastes of the founding fathers are likewise obscure.

"Your two might have been able to dance a bit. They might have known songs. They may well have had a whole repertory of good old songs. They may even have had some in Irish."

This is Reg, but I suspect he's being generous to his old friend Bill. Patrick and Thomas were not peasant farmers; they were urbanites. Ennis was town rather than country (as were Limerick and Kilkee). Reg, in *A Few Tunes of Good Music*, comes up with a succinct reason why Patrick and Thomas, and thousands of their compatriots, should culturally uproot themselves:

> ...the stimulation was not forthcoming and the basic issue of survival in an alien setting dominated motivation and sapped energy.[8]

Reg and I won't admit the possibility that Patrick and Thomas were fans of music hall. Music hall is the elephant in the room. Bert Lloyd, interviewed in 1970, reckoned "folk song in its classical form has certainly not been the music, the great popular music, of the industrial proletariat in England."[9] The matter is one of deep concern.

"What is English traditional song?" asks Reg. "I think about it. Tony Engle thinks about it all the time. How much of it was around? How common was it? We've got all sorts of prejudiced views that everybody was singing 'A Sailor Cut Down in His Prime', but, you know, were they?"

(Interesting example. 'A Sailor Cut Down in His Prime' is the root song of 'St James' Infirmary' and 'The Streets of Laredo', so the song was in the repertoire of African-American songsters and cowboys. It's

8 Reg Hall, 'The Transplantation, Survival and Adaption of Irish Rural Music and Dance in London', *A Few Tunes of Good Music*, p. 25.

9 Quoted in Roy Palmer, *Working Songs*, op. cit., p. iv.

a foregone conclusion: a lot of people were singing 'A Sailor Cut Down in His Prime'.)

When the Leader brothers arrived in London in the 1850s, raucous anarchy prevailed on the forecourts and back gardens of large pubs. 'Gleemen' (an ancient word) would buttonhole patrons as they sat on long trestle tables and entertain with a song or comic patter for tips, while waiters provided more misrule than nourishment. The contact was direct and upfront. The star system dates from a decade or so later, when publicans started to pay performers and charge admission. The resulting cash-flow generated the capital to build purpose-built places of entertainment or refurbish existing pubs on the music hall model. The Bedford arose on the site of the tea garden of The Bedford Arms in Camden Town, and The Metropolitan on Edgware Road was originally The White Lion pub.[10] Their names were redolent of opulence, pleasure and manifest destiny: The *Palace* Theatre, Holborn *Empire*, Victoria *Palace*, The *Empire* Theatre of Varieties; The Finsbury Park *Empire*. The direction shifted from one-to-one revelry to passive, packaged hilarity and, finally, to worship from afar.

And although the Victorian working class were downtrodden and exploited, they didn't act like it at leisure. Music hall songs are primed for generosity and laughter: some, like 'Champagne Charlie', conjure a world of infinite hospitality. This conscious flight from reality, together with all those Palaces and Empires, made left-wing thinkers wary of music hall. Others grasped its irrational appeal. Vladimir Lenin, exiled in London, nailed it in a letter to his friend Leon Trotsky in 1902: "It's the expression of a certain satirical attitude towards generally accepted ideas, to turn them inside out, to distort them, to show the arbitrariness of the usual." [11]

Ewan MacColl was wary. John Foreman, with music hall in his blood, chuckles about the time he put one over on MacColl.

"There was a feeling that music hall songs were imperialistic; they were part of a period of history when the world was all wrong. There was something *wrong* about the music hall. I remember, as a political statement, when Ewan MacColl let me sing at his club – because he had a marvellous club, you know, Ballads and Blues was one of the best

10 John Alexander, *Tearing Tickets Twice Nightly: The Last Days of Variety* (Arcady, 2002).

11 Letter from Lenin to Trotsky, quoted by Victor Sebastian in *The British Road to Bolshevism*, BBC R4, October 18, 2017.

clubs there was – he let me sing, and I sang 'Slap Dab' [chuckles], about wallpaper [more chuckles], a music hall song."[12]

Social realism was not on the bill. Magic realism, although the term had not yet been coined, was the object. Consider a Foreman favourite, a song associated with Gus Elen ("a great man" says Bill).[13] Elen sung 'If It Wasn't for the 'Ouses in Between' in the persona of a market trader, or costermonger (it was recorded on February 21, 1899). Confronted with a yard full of rotting vegetables, Elen sees a pretty market garden. Simple props are enlisted to change urban squalor to rustic idyll: beetles in a bucket substitute for bees in a beehive, and imitation horns aid the metamorphosis of a scraggy donkey into a contented cow. More elaborately, a pulley and a rope attached to the chimney are needed for a full appreciation of the similarity between the municipal gas works and the contours of an Alpine mountain range. The hard-boiled patrons of the halls wholeheartedly endorsed such whimsicality. They used the same methods to turn Beckton Gas Works into Rhubarb Junction.

> Not only did Canning Town have the world's largest gas works near-by, poisoning the area, it was also close to the London docks and Silvertown, a home for chemical works; a munitions factory, which blew up in the first world war; a soap factory; the Rubber Tyre, Gutta Percha and Telegraph Cable Company, where my mother clerked for a while after she left school; dumps for storing manure; three vast flour mills and two huge sugar refineries. Silvertown has frontage on the river Thames. To its north, Canning Town, lying to east of where the River Lea runs into the Thames, had been marshland and pretty inaccessible until a decent bridge had been put across the Lea in the early 1800s. To my mother, born in 1896, London proper started on the other side of the Lea's iron bridge; Canning Town was proudly Essex. A passing stranger might well have been perplexed by the distinction. To the casual eye it was one appalling stretch of docks and deprivation all along the river from Wapping, just east of the Tower of London, down to Galleons Reach.[14]

12 John Foreman, interview with the author, March 21, 2014.

13 Bill Leader, interview with the author, April 10, 2013. Bill Snr was a big fan too.

14 From *If I Remember Writely – The Unsound Memories of a Soundman*, an unpublished memoir by Bill Leader.

Charles Dickens, however, believed in calling a spade a spade.

> Canning Town is the child of the Victoria Docks. Many select such a dwelling place because they are already debased below the point of enmity to filth; poorer labourers live there, because they cannot afford to go farther, and there become debased.[15]

Chas was a great speaker and a popular hero, but if he had toured the halls with this kind of material, he would have been pelted with rotten tomatoes. Jo takes up the cudgels on behalf of 58 Lawrence Street, Granny Leader's home in Canning Town, in the middle of a long terrace:

"People talk about that area as being slum accommodation, but it wasn't. Granny Leader's house was quite a nice house. It was an odd shape, because of the bend in the road. The front of the house looked like any of the others: front door, a window at the side and a window above. And when you went in there was a passageway and a staircase up, and you went round a little corner to the door in the back room. But then the back room curved with the bend. It was a very big back room. And beyond it, down two steps, there was a kitchen, with a big cast-iron range."[16]

"It always seemed to me to be a big house," Bill remembers, "because I came out of a council house which was economically designed. They didn't have semi's in those days. Waste of bloody space. So it was a long terrace of houses. The rooms were big and high. There were two main rooms on the ground-floor, and there was a kitchen and a scullery going out into the yard. Oh, the front door was on the pavement, straight out onto the street, but there was a big old yard at the back. And everyone's was like that."[17]

Yes, but without the kink.

---

15 Charles Dickens, *Household Words* Vol. XVI, 'Londoners Over the Border' (1857), cit. https://en.wikipedia.org/wiki/Canning_Town.

16 Jo Pye, interview with the author, December 11, 2013.

17 Bill Leader, interview with the author, October 29, 2015. "Then to the left-hand side of both rooms, there was a passageway, and it ran down from the front-door, down the side of the two main rooms, down a few steps into the kitchen and then into a scullery, that had an old sink and a boiler, that you fired up. A copper type thing. It was a water heater. And I think, again from Jo, that one of the girls would get a bath in this."

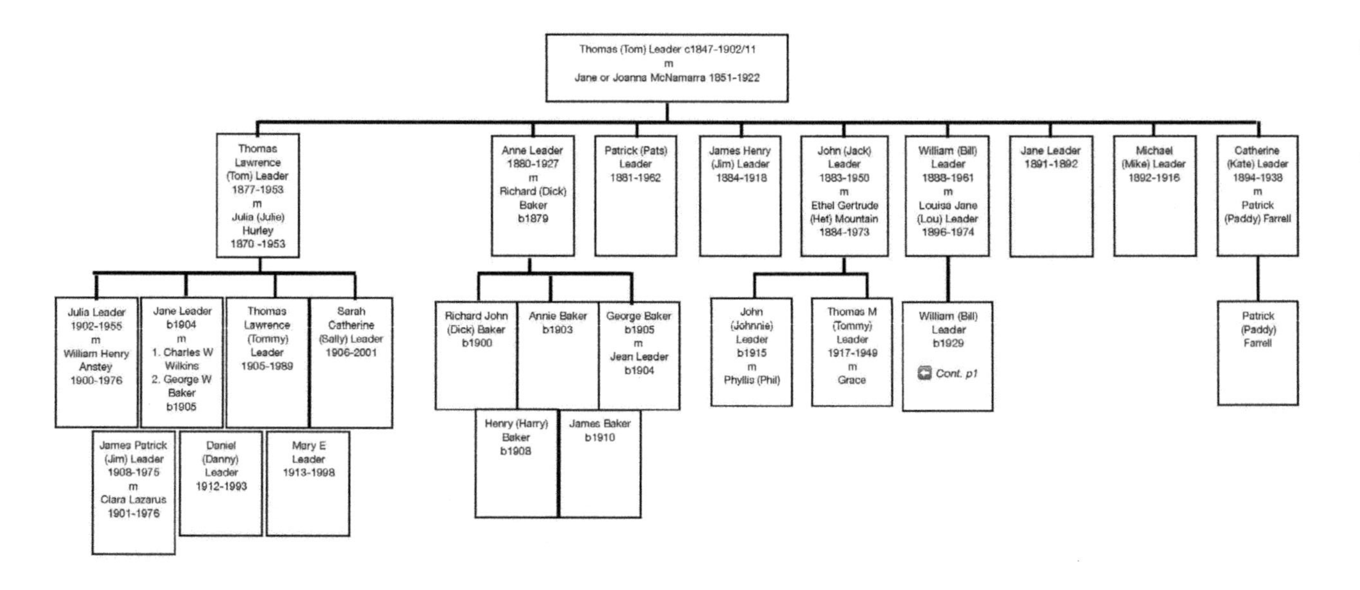

Fig. 2 *Bill's paternal aunts, uncles and cousins*

"It had an electric cooker and an electric light," continues Jo. "And the scullery, when they pulled it down in 1962, still had a built-in copper. You kept a permanent fire under there and you could have constant hot water. One of my mum's sisters used to light it up, get in and have a bath. You had an outside lavatory. You didn't have a bathroom, but at least they had their own lavatories. My mother [Margaret] was absolutely horrified when she came to live with Aunt Lou, and she went to Leeds and went through Armley on the bus. Back-to-back houses with the washing hanging in the street! 'How common!' She said, 'And they call Canning Town common!'"

The 1891 census lists Bridget, a.k.a. Great Granny Leader, as a greengrocer in Rathbone Street, a Canning Town street market. Family legend has her as the proprietress of a flower shop. Her husband was the Patrick who came from Ireland in the 1850s, who died on December 22 1877, at the family home of 37 Denmark Terrace, East Ham. His death certificate gives the cause as enlarged liver and peritonitis; in other words, cancer of the liver. He was 38 and had worked as a stoker since his arrival in England. He had three children with Bridget – Thomas, Patrick and Bridget (the younger). Thomas, the eldest, is the grandfather Bill and Jo have in common. Bridget remarried and became Bridget Gaimon, although, strangely, she seems always to have been known as 'Great Granny Leader' or 'old Mrs Leader'. With a temper to match her flaming red hair, and with piercing blue eyes, Bridget embodied a certain type of Irish matron. She died in Whipps Cross at Leytonstone in 1916.

"Even when I was a child, people didn't want to go to Whipps Cross. It had been both a workhouse and isolation hospital and had the reputation of a place where once you went in you wouldn't come out."[18]

The address given on Great Granny Leader's death certificate is 110 Croydon Road, which is the home of second son Patrick, and daughter-in-law Agnes. Patrick was comparatively well-off, says Jo. His private grave in St Patrick's Roman Catholic Cemetery in Leytonstone cost eight guineas and is still there, whereas Bridget's grave in the same cemetery is public and has been used dozens of times. He left £339 and sixpence, "which was a hell of a lot of money in 1933. Perhaps he was a good card player like Aunt Nance. I don't know, I have no idea. They just lived in an ordinary house like everybody else, because

18 Jo Pye, email to the author, March 29, 2018, and Thomas as a road sweeper, below.

Bridget Gaimon, or Great Granny Leader, c. 1900 (collection: Jo Pye)

He was a large and imposing man… Thomas Leader a.k.a. Grandad Leader

Louisa Martha and Thomas (front) with Harry and Nance Davis

Thomas with Lily and Patricia (cousins)

I remember being taken around to see Aunt Ag when I was a little girl… He worked in the gas works."

Bridget and Patrick had a small family by Victorian standards because Patrick was dead at 38. His brother Thomas, who married Jane/Joanna McNamarra in 1873, had nine children, of which the sixth was Bill Snr.

> Dad came from Battersea, from that part of London where the pokey houses and cramped streets nestled amongst the arches and bridges that carried the railway lines into Victoria or Waterloo stations. The street and the little houses have gone now: the railway remains. Dad was at the younger end of nine children: Tom, Anne, Patrick, Jack, Jim, Bill, Mike and Kate. Most of them customary Irish names that any good Catholic family from County Clare might resort to; except for Bill. Why Orange-tinged William?[19]

Census returns are vaguer about deaths than births. Thomas was alive in 1901 but he is absent from the 1911 census, so presumably he died in the interim. His place of birth is given as Clare, which is a blow for the Limerick lobby. He was aged 57 in 1901 and worked as a 'labourer in a gas works'. Jo's research on the Ancestry website, where a lot of workhouse records are available, shows that Thomas was admitted to Ward B2 of the Lambeth Workhouse on Tuesday April 15, 1902, and again on July 28 the same year.

"And on that occasion, it says that he came from the relief ward at Stocksbridge, but I don't know what that means. His age was given as 59 and then it says that he released himself – 'released by his own request' – on the first occasion 'on Thursday May 1st, 1902', and, on the second, 'on Saturday, August 16, 1902'. So, he may have felt better and then relapsed. Workhouses were being used as hospitals at that time. It's got to be him because the address on each occasion is given as 15 Brooklyn Street, Wandsworth, which is the address on the census the year before, and his wife's name is given as Jane. To have gone into hospital on two occasions within a short time span in the same year does rather suggest that he was seriously ill. I haven't found any death records for him yet."

19 *If I Remember Writely*, ibid.

Circumstantial evidence points to 1902 as the likeliest year of death for the second of the founding Leaders.

Jane, aged 61, lived at 69 Brooklands Street, Stockwell in 1911. She is listed as the head of the household.

"I don't know if I've written this down wrongly, but I've got her down as an electric railway labourer! Now whether she was a cleaner, I don't know. She's listed as being born in Lambeth. She was always listed as being born in England, which is not what Sally said."

Official records, contrary to family legend, indicate that Jane Leader, née McNamarra, was English born.

> My grandfather worked at Beckton Gas Works until, towards the end of his working life, he got a job in the open air, navvying. Even later on in his life, when I came to know him, he still had the remains of the physique of a man who had laboured hard all of his life.[20]

Was the grandfather Bill and Jo had in common a 'navvy'? Bill remembers him digging up roads and laying down tram tracks. Jo doesn't think so.

"Dad said that when he left the gas works, he went to work for the council. I seem to remember that he was a road sweeper, but I could be wrong about that."

The 1891-1911 census returns for Thomas, born in 1865, and the birth and death certificates of his children, agree that he was a gas stoker, sometimes a 'gas labourer'. The 1911 census specifies Beckton Gas Works.

"My grandfather on my mother's side spent most of his time working in Beckton Gas Works," Bill informs. "You were lucky to get out of one of those places alive."[21]

He was a large and imposing man, toughened rather than worn by heavy manual labour, with unusually delicate hands. His long, tapering fingers were so sensitive that, blind in his final years, he could identify playing cards by touch and join in the family's favourite pastime. Or that's what Maggie, Jo's mother, told Jo. She has no personal recollection because he died, in August 1942, two days after she was born. ("Everyone's a critic!") Bill was eleven and remembers him well.

---

20 Ibid.

21 Bill Leader, interview with the author, March 2, 2012.

"He was a big strong sod. He always struck me as being very old, but I was very young at the time. (He wasn't as old as me now, I guess.) Liked his drop of beer. To me he was the patriarch. He sat in a corner in a big chair, next to the radio, and listened to the news and the football results. It used to be a Saturday by the time we arrived for whatever was going on. This was in Canning Town."[22]

Bill isn't aware of any particular interest in music.

Pause awhile and consider the dates of James Henry and Michael in Fig. 2, above, that twig of the family tree on which Bill Snr and his siblings sit. In the 1911 census Michael's occupation is given as a 'bottler in a brewery'. Charrington Brewery, situated in Bethnal Green since the early eighteenth century, is the likeliest employer. Within a few years lives would be thrown away wholesale, and nobody much missed a bottler in a brewery.

"I never knew anything about them. Nobody ever mentioned them," says Jo. "I only knew the living ones. I knew that Uncle Bill had a sister who died. That was Kate. She died quite young. They talked about her, but they didn't talk about the two brothers. I only found them because I was looking for my mother's family. I thought, oh, let's see if there are any Leaders. There's quite a lot of information now, an increasing amount of information; originally you just got the name of the person, if they died. You don't get any information if they didn't die. Or rarely. You do for the Merchant Navy. Uncle Harry was extraordinarily lucky. He was allowed to leave his ship and go somewhere else about a month before she was sunk. I did know about Uncle Pat, Uncle Bill's brother. He was in the Royal Navy. My dad told me that Uncle Pat was a submariner, but I've not found any evidence of that. I've found a lot of information about the ships that he did serve on, but none of them were submarines. He was a stoker, but aboard ship, not in the gas works. He lived to be God knows how old." (This Pat died, unmarried, in 1962, aged eighty-one.)

"I've subsequently found Michael. We went to Thiepval on the centenary, so 2014. James, the other one, is at Le Tréport, which is a long way away, off the beaten track. He is actually buried. Michael was never found. That's why he's commemorated at Thiepval, because no one there actually has a grave. It's that enormous Lutyens redbrick

22 Bill Leader, interview with the author, October 29, 2015.

structure. It's got thousands and thousands of names on. It's not quite as big as the one at Ypres, the Menin Gate, but it is huge.

"They were both in the London Regiment, but the London Regiment was enormous. Practically the entire male population of London was in some battalion of the London Regiment. And like everybody else they got the Victory and the British Medal. Everybody got one for turning up.

"Interestingly Michael went AWOL at one point. He's listed as being AWOL from 11pm on the 16th June 1916 to 4am on the 19th June 1916. He forfeited fourteen days' pay. He was killed three months later. I don't know how he got killed. It may have been a sniper, and nobody knew. It's like seamen. They stop your money the minute your ship goes down. And in some instances, they stop payments to the families. His mother was given outstanding pay of one pound twelve and ninepence, in 1917. In 1919 she got another three quid. The war gratuity varied. People in the same family got three times as much as other people got. A friend of mine is a military historian, and he said it was quite arbitrary, a lot of it. Some people got a lot of money, and some people got fourpence if they were lucky. His mother received twenty-two pounds seven and ninepence for his brother James, including nineteen pounds war gratuity, in April 1919. He died of wounds. He wasn't killed on the battlefield. That's why he got a burial. I don't know if that's why he got more money than his brother did."

Bill Snr was out of harm's way because he was in a reserved occupation.

"If you were working to make the shells to kill the other bastards, you were in a reserved occupation," Bill says.[23]

23 Bill Leader, interview with the author, November 6, 2015.

If there is such a thing as truth it is as
intricate and hidden as a crown of feathers.

ISAAC BASHEVIS SINGER, FROM THE STORY 'A CROWN OF FEATHERS'

II

Other Chimney Corners

THE CHILDREN STRADDLED THE old and new centuries and were so numerous that patterns of hair colour began to repeat, like geological strata. The Victorians, Nance and Lou had black hair; Molly was auburn like Harry and Patrick who followed her (it was remarkable how alike the two boys looked). The Edwardians, Lil, Ted and Jim, reverted to black. Would Kit, born in 1914, count as Georgian? The whole crew had blue eyes except for Pat, whose eyes were green, and all had long, tapering fingers, a distinctive family trait. Two didn't make it: Walter Thomas, the first-born in 1891, lived only days, and his brother Thomas, born in October 1894, died before his first birthday.

> My mother was the fourth-born of ten but led her life as the second eldest of eight: infant mortality being what it was. Walter, Anne, Thomas, Louisa, Molly, Harry, Pat, Lil, Ted, Kit. Louisa was her mother's name; Walter was her

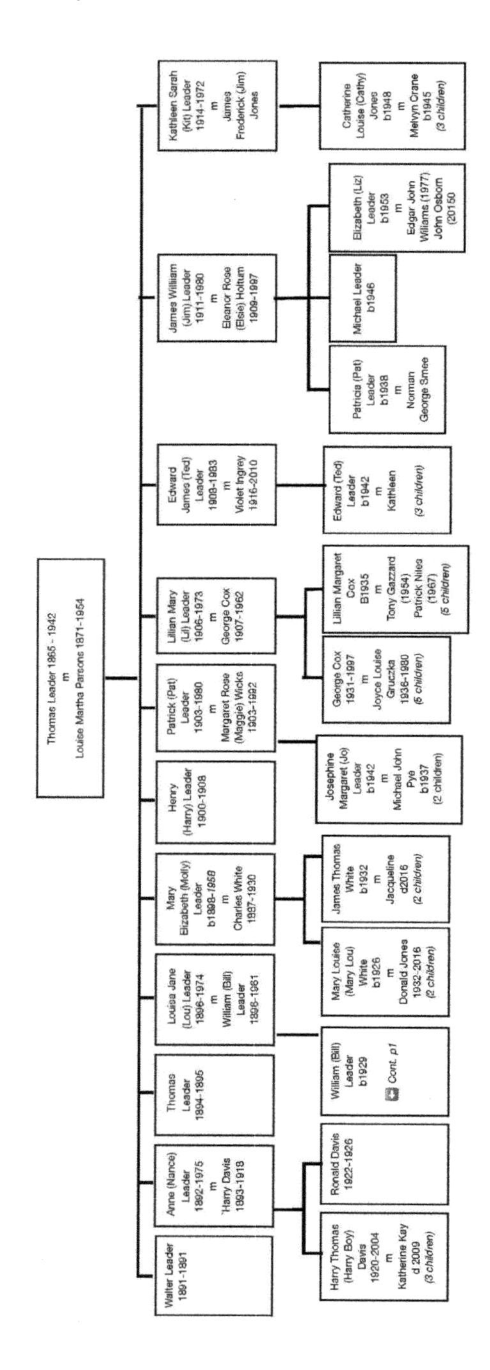

Fig. 3 *Bill's maternal aunts' uncles and cousins*

> mother's brother's name. Her mother was one generation away from the English rural Midlands, with no Celtic connection that I have discovered: the only non-Irish person in my lineage.[1]

The advantage of having a big family, and such abundance was commonplace then, was that children could be relied upon to entertain themselves. In this way eight were less trouble than one, although child rearing in quantity has its challenges. It goes hard on the eldest, for example, to be routinely expected to child-mind, despite being only slightly less small than her charges. Nance, an impetuous child, once abandoned her siblings and followed a barrel-organ around the streets, skipping all the way. She was never allowed to forget the escapade. Emigration to the States offered no protection from family folklore. On a visit to the old country Jo met her with, "Oh Aunt Nance, you're the one who dances after barrel-organs."

"Who told you that? Our Lou?" asked Nance sharply.

Yes, except Jo's father, Patrick, had told her the same story.

Whereas Lou and the street singer was a confidence, meant for Gloria alone. Gloria lived with her parents-in-law after marriage to Bill, and with Bill away on National Service and Bill Snr fully occupied between work and union duties, the two women spent a lot of time together. Lou told Gloria things she never told anybody else. It might be worth setting down Gloria's complete and unvarnished account of Lou and the street singer:

"They had shutters in London, in the area… West Ham. Long rows of houses. And they had wooden shutters. And at night, they didn't have curtains, they came out and put the shutters up. And she said they'd be a street singer coming down the street. And she used to then take a long time putting up the shutters. And by the time the singer had gone, she'd learned the song."[2]

I can picture the scene. The sky is full of sunset. It is 1913 or thereabouts. Nothing bad has happened (that is, a generation had not yet been sent to slaughter). Quite miraculously, none of her brothers and sisters are around as Lou puts up the shutters of the home in Canning Town. It is twilight and still, with shadows lengthening on

1 *If I Remember Writely – The Unsound Memories of a Soundman*, an unpublished memoir by Bill Leader.

2 Gloria Dallas, interview with the author, February 27, 2013.

the deserted street. The one clear sound, the voice of a passing street singer, is magnified in the calm, and Lou fights a Nance-like impulse to follow the siren song.

Gloria omits to mention its name. Possibly she never knew. I fancy some parlour ballad of unrequited love; shall we say…

Go, pretty rose, go to my fair
Go tell her all I fain would dare…[3]

Looking neither left nor right, the singer affects not to notice Lou as he draws nearer. There is something a bit self-conscious in his distraction as he passes without a sign and then disappears at the same sedate pace. His strong, melodious voice lingers long enough for Lou to memorise the song. When the last echo has faded, she tries it out softly, as if handling some fragile gift.

Oh, were I like thee, so fair a thing…

This idyll of old Canning Town is possibly nothing more than a girlish, commonplace reverie. I feel guilty about grafting an added layer of romance on Lou's lily.

"This is a lovely story about Aunt Lou putting the shutters up and the song and everything," comments Jo, responding to an earlier draft. "But it's just a story. The houses didn't have shutters. I know the houses didn't have shutters because Aunt Lou lived on Clarkson Street and I used to go to Clarkson Street School when I was a child. I know those houses didn't have shutters because I knew those houses. But if she worked in a shop…"[4]

Of course. Shops have shutters, not houses.

"Certainly not in London. And certainly not the kind you would have to bring out from the house and attach. That story could come from when she was older and working at Judd's shop. It certainly has nothing to do with where she lived. There were no houses with shutters in Canning Town."

Yet Lou and the street singer live in Gloria's mind, and they have taken up residence in mine. At the very least, Nance and the barrel-organ and Lou and the street singer indicate a genetic susceptibility to song. I might add that Lou was an inveterate habitué of the cheap seats

3 'Go Pretty Rose' by Theo Marzials.

4 Direct quotes from Jo Pye in this chapter are culled from two interviews with Bill's cousin: on December 11, 2013, and March 20, 2018.

of the Savoy theatre, home to the D'Oyly Carte company, the exclusive producers of Gilbert and Sullivan operas. In contrast to the later D'Oyly Carte policy of Gilbert and Sullivan for all, in the nineteen-tens and twenties the management actively discouraged a working-class audience by charging high admission, having a strict postal booking system and scheduling start times before the end of the working day. Lou rose above it all, and the celebrated electric recordings – the 1926 *Mikado*, with Henry Lytton, the 1927 *Gondoliers*, still enjoying good sales on CD, and the 1927 *Trial* – formed the soundtrack to Bill's childhood.

Yet those in the know insist that, in the matter of recreation, music came a poor second to cards in the Leader household.

"Before you finished your tea, the cloth would be off the table and the cards were out," says Jo. "Aunt Nance would get quite frustrated if she was there, because you had to play a simple game if some of the children wanted to join in. My mother [Maggie] was not remotely interested in playing cards and nor was my Aunt Violet. Aunt Violet could just about manage Newmarket. So we used to play Newmarket, which Aunt Nance used to find very frustrating. And then after a while the kids would get bored, and my mum would get bored and Aunt Violet would get bored and then, *here come the real cards!* Aunt Nance and my dad and Uncle George, who had been thirty years in the Royal Navy, and Uncle Jim who had been in the army, spent a lot of time playing cards."

George, Lil and George's son, ten months younger than Bill, was a chronic gambler who ran up appalling debts. Nance, on the other hand, was a champion at pinochle.

"She and her husband would sit and play pinochle pretty well all day," remembers Bill.[5]

"She financed a trip back from America on her winnings, she was that serious," corroborates Jo, admiringly. Lou, for her part, excelled at whist, and abstained from bridge.

"We played working-class card games, not middle-class card games."[6]

The children were raised as Catholics. Their father, Thomas, was Irish and a cradle Catholic, and their mother, Louisa Martha, was a convert. This made her feel her religion more keenly. Louisa Martha

5 Bill Leader, interview with the author, April 10, 2013.

6 Bill Leader, interview with the author, September 25, 2015.

actively pushed the boys to become choirboys and altar-boys.

"The thought of my dad with his bright red hair and a red cassock!" says Jo, smiling at the picture.

Pat, Jo's father, was never known to sing in public after his time as a chorister at St Margaret's, but he could be euphonious about the house, and was not in the least concerned, as a firm atheist, about the sacred bias that had crept into his repertoire.

The relationship between the Leaders and their god is interesting. Kit, the youngest, tended to gabble Christianity's second most popular prayer as "Our Mary full of gravy one two three" and, in the Apostles' Creed, where Jesus suffered under Pontius Pilate, Kit would have it as "he sat upon a bunch of spiders."[7] Molly grew into conventional piety once in the States and became a Daughter of the British Empire and an American Episcopalian, or vice versa.

"And our Mary, her daughter, is actually quite religious."

Jo ponders and qualifies this last remark.

"Well, she goes to church."

In later life Aunt Nance embraced Catholicism. This too is overstating it. Let's say she observed the formalities of the faith to please her Catholic daughter-in-law.

In truth, the regime at St Margaret's, the small school attached to the small church on Barking Road,[8] was so authoritarian that it produced a counter-reaction. If the religious indoctrination was supposed to result in the submission of the children to a higher authority, it had the effect, in Lou's case at least, of instilling devotion to the higher authority of Marxism.

"It was six of the best if you hadn't got a clean hankie, and six of the best if you hadn't clean shoes. They had a priest called Malarky. I always thought my Uncle Jim was taking the mickey, because he used to call him Father Malarky – 'malarky' is a Cockney term for mucking about – but actually that was his real name. I thought his name must be Malachi, which is a fairly common Irish name, and it was mispronounced. No, his name actually was Malarky, because I've been there since to look at the records. It is spelt M.A.L.A.R.K.Y.

---

7 Bill Leader, annotation, August 25, 2018.

8 St Margaret's and All Saints Roman Catholic School was opened in 1860 with a building grant made by the Poor Schools Committee. It closed in 1940, after being wrecked by bombing.

"Aunt Lou was a sort of a prefect. She and Aunt Nance, so one of them told me, I can't remember which, used to hand out the shoes to the poorer kids who didn't have shoes. Aunt Molly was, I think, embarrassed by her two older sisters. Aunt Nance and Aunt Lou had a reputation for barging in if they thought it was necessary to do so."

A priest from St Margaret's called on Thomas and Louisa Martha and tried to persuade them to let Lou stay on at school. She was a bright girl and had a good head for learning, he said, and dangled the prospect of a scholarship. He predicted a bright future for Lou, perhaps rising to the heights of headmistress. But to fulfil her potential career she must stay on at school. Thomas and Louisa Martha, much impressed, gave the matter careful consideration. The upshot was that Lou left school aged 13 and went into service.

Doubtless the same discussion took place in the households of many bright working-class girls. It forms a sub-plot in Winifred Holtby's *South Riding*, one of Lou's favourite reads.

Tom,[9] Bill's son, was enraptured by his grandmother's stories.

"She was an amazing woman and an excellent grandmother. Full of stories. And because she'd done so much, she was good at proper stories. They weren't just truisms or whatever, they were her experiences. And they were said with feeling. It was very educational. I think she had that effect on everybody. She did impress a lot of people or had an effect on a lot of people. A fantastic cook as well. All this stuff about the Victorians not being able to cook. She bloody could. She was great. Soggy cabbage? Never heard of it!"[10]

"When you went into her flat, you'd often come across her sitting in the armchair, with her ankles crossed, reading," says Gloria.

My Lady of the Chimney Corner, by Alexander Irvine, was Lou's touchstone.

"Apparently she read it several times," says Gloria.[11]

"She always had a copy of that. If she lost one, she would buy another. Although she didn't lose many, she kept them close by her," confirms Bill.[12]

---

9 Tom is unique among the Leaders for being a birth Tom, as opposed to a shortened Thomas, but his own son, Thomas – a.k.a. Tom – reverted to type.

10 Tom Leader, interview with the author, March 6, 2013.

11 Gloria Dallas, interview with the author on August 20, 2016.

12 Bill Leader, interview with the author, April 10, 2013.

"I've got her favourite copy downstairs. She gave it to me. Bound in red leather," says Jo.

"There's another story that she told me," continues Jo, exhibiting a scepticism that has not yet dawned on Tom, "and I thought this was a wonderful story, supposedly about my grandparents, Thomas and Louisa Martha. Granny is supposed to have got them a meal and fed the children, but there wasn't enough for her. But she kept the meal for Thomas, because he was working. And then when he came in, he was drunk and he went to bed, and so she ate the food, but she dipped her finger in the gravy and wiped it around his lips, so when he woke up he would think that he had had the meal. Unfortunately, I've read that book too. Or at least my mum had."

She thinks it might be *My Lady of the Chimney Corner*, but I suspect it is not, from my own reading.

"So off she went," Tom continues, recalling Lou's early days in service. "She was very miserable. Friends came around one Sunday morning and she had been left with everything to do. The family had all buggered off. And they just said, 'Right, that's it, you're coming and staying with us. We'll sort it out with your parents.' I don't think she ever told me what 'sort it out with your parents' meant, but it must have been some sort of financial thing. And she went and lived with them. They had a hardware shop, I seem to remember. She lived in a room above. And either the room she was in, or one room in the house, had three walls that were just bookcases, and it was all fairly left-wing stuff, and while she was there, she read the room. That was her education."

Lou's eight-year-old grandson made an uncritical audience but even he was puzzled by that "we'll sort it out with your parents." Clearly, Lou was talking about the Judds. Once again, Jo delivers the relevant data, gleaned from the 1911 census, which lists Lou as a servant of the Judd family. Lou had left school the previous year.

"Hang on, here we are. 1911 census – '155 Earls Court Road: Bartlett Judd'. The head. He was sixty. 'Oil, colours, china and glass dealer. Mary Judd, his wife, 32.' So half his age. Maybe that's why she was so friendly with her servant."

Mary Judd was a suffragette and more progressive than your average Edwardian employer. She was fond of her live-in servant and took her to places that would normally be off-limits to a working-class girl from the docklands.

"They used to go to meetings of the Theosophical Society at Red Lion Square and things."[13]

Lou was free to read anything she liked from the library above the shop. She made a stab at *The Communist Manifesto*. The author, Karl Marx, proposed practical ways to make life better for ordinary people. The other person who tried this, she was aware from St Margaret's, was Jesus. And although the success of Christ's mission had been drummed into her by the priest teachers of St Margaret's, her own experience taught a different lesson; it was something she was still struggling to articulate, something about being a member of a social class that had sunk below the point of enmity, living in material conditions that needed fixing. Might Marx succeed where Christ failed? A light began to glimmer in her soul. She devoured texts by Friedrich Engels, Ferdinand Lassalle and Pierre-Joseph Proudhon. It was a strange language, yet she understood it.[14]

Did the job at the bakery come before or after the education of the Judds' reading-room? After, I think, for now she could see that rejuvenating stale cake and selling it to your customers as fresh-baked was a clear case of social antagonism, dictated by the laws of capitalist production.

'What book did you read that in?' taunted Nance.

Das Kapital by Karl Marx replied Lou with pride.

Theosophy! Mrs Judd was impressed by Madame Blavatsky's profound philosophy, but to Lou it was nonsense. Lou placed as little faith in cosmic intervention as she did in the divine kind. The Substance-Principle had the whiff of stale cake about it. Lou was in favour of direct action, and so should Mrs Judd be, as a good Suffragist. This was delivered with characteristic forthrightness, and Mrs Judd was charmed.

A few notes of caution here. Conway Hall in Red Lion Square, where Mrs Judd and Lou allegedly went to the theosophy lecture, was built in 1928 and opened to the public in 1929. Lou had been in the States a full five years by 1929. And Mary Judd is listed in the 1911 census as 'wife and housekeeper' (there is a son, Harold Judd, aged ten). This is unusual.

---

13 Bill Leader, interview with the author, October 29, 2015.

14 "I think you should mention that turning Mum into a one-woman Marxist study group does have an element of fantasy." Bill Leader, annotation, June 24, 2020.

"If a female member of the family is listed as a housekeeper that's either because she's a paid servant or because she's an unmarried daughter. You don't often get the wife listed that way," Jo informs.

There goes my image of Mrs Judd as a lady of leisure, taking her servant to self-improving lectures to keep her company.

And, still in the grey area between romance and reality, what are we to make of brother Jim's story about Lou agitating on behalf of the office-staff of the India Rubber, Gutta Percha and Telegraph Cable Company? The business was known as Silver's, after Samuel Winkworth Silver, the largest employer in Silvertown. The shop floor had been unionised and Lou was organising representation for the white-collar workers. According to Jim, she got herself and brother Pat sacked for her efforts.

"How true that is, I don't know, because in 1911 my dad was only eight, and Uncle Jim must have been all of five," says Jo, striking a sceptical note.

Bill also thinks it unlikely.

"She's never mentioned that to me, and she would have done, I'm sure. My mum was a great narrator, and she would have mentioned that particular event and reiterated it many times."

Lou's reading confirmed that the prevailing order was not inevitable or beyond repair. She turned her face towards the sun, which oddly shone from the direction of the USA. 'Oddly' because Lou had mixed feelings about the USA, source of the most rapacious capitalist system in the world. Nance had gone out there in 1923. Lou followed in 1924.

"She would have been 28 years and three months by the time she got there," says Jo. "She travelled steerage, but steerage on the *Leviathan* wasn't as awful as some of the immigrant ships. The *Leviathan* was built as a luxury-liner, and steerage on what was originally a luxury-liner wasn't so bad. (Just as third-class on the *Titanic* was not the awful thing you get in that dreadful film.) It was actually better than first-class on some smaller ships. I mean you shared facilities, *but you had facilities*. That was something. I think all the world and his wife must have gone to America on the *Leviathan*."

Jo, erudite and articulate, has once again saved me the trouble of first-hand research.

"The *Leviathan* was built in the First World War for a German cruise-line company, The Hamburg America Line, and after the war the Yanks commandeered her, as part of the reparations, and Woodrow Wilson renamed her the *Leviathan*. I think she was called *Kaiser Wilhelm II* to start with [*SS Vaterland*, in fact]. And she spent a lot of her time as an immigrant ship ferrying backwards and forwards across the Atlantic."

The outbound journey went from Southampton to New York via Cherbourg. Jo paraphrases the famous line engraved at the foot of Liberty as "Bring me your whatever and we'll squash them even further" and it strikes me with renewed force that Jo is worth her weight in gold. I imagine the immigrants as far too tired, poor and huddled to raise a whimper, much less a cheer, as the ship sailed past the Statue of Liberty in New York Harbour. I might be mistaken. I need to pin up Bill's words where I can see them:

Nothing truthful has ever been written. It starts as being your point of view, as far as you can be truthful within your own perception and knowledge, which is limited.[15]

We do have some hard facts, however. The date was October 17, 1924, and Louisa's immigration visa was number 2488. These details are in her passport, which survives. "The validity of this passport expires July 14, 1926," runs the legend. The passport was renewed and given a new expiry date of July 14, 1931, but this was not needed. By July 14, 1931 Lou was back in the UK and the passport was superfluous. She never crossed another international boundary in her long life.

She had 25 dollars in her pocket at Ellis Island, the minimum required to avoid immediate repatriation.

"She said the way they treated you at Ellis Island was really awful," says Jo. "Her nationality was English. She was able to read and write. English speakers were treated better than a lot of East Europeans and Jews, who didn't speak English. And the next of kin is given as her father. And her occupation is given as a domestic. And her final destination – this is the bit I love – was to her sister, in Brooklyn, 'for an indefinite period of stay.' So she didn't intend to come back. And she said 'no' to all the political questions, which I think is wonderful. Then it gives her immigration visa [number 2488]. The Ellis Island site is actually quite good, although it's not as good as it was. I think it's gone down a bit."

Today Ellis Island is occupied by the cavernous Museum of Immigration. Visitors are routinely exhorted to have a nice day and

15 Bill Leader, interview with the author, September 15, 2017.

directed to the Museum Shop with its Statue of Liberty Barbie doll. The Museum of Immigration promises to connect American citizens to their immigrant ancestors and have fifty-one million entries in its database. The places where the huddled masses massed in huddles are quiet now, and open, clean and sterile.

She had 25 dollars, the minimum required to avoid swift repatriation.

Why do we read? It's a pain-free way of getting out of our heads and into someone else's. Books demolish social differences and demonstrate a common humanity. And books like those in the wonderful library above the china and glass shop advance the cause of truth and knowledge. Lou, an avid bibliophile, read to find something alike in others. It may be that an acquaintance with the works of Henry James primed Lou for Judge William Caldwell Coleman, and her reading of *Black April* by Julia Peterkin (another favourite book) raised her awareness of the inner lives of his domestic staff. Peterkin was a white author praised by Langston Hughes for her non-stereotypical portrayal of African American characters. Lou's tastes leaned more to Peterkin than James: she was drawn to those with too little money rather than too much.

"She used to read and re-read Upton Sinclair. *Oil*, I think was quite famous. Have you come across that?" [Only as the Coen brothers' film *There Will Be Blood*, which I've not seen.] "And *The Jungle* is all about the Chicago canning factories, and the exploitation of workers, mainly immigrants, working under appalling conditions, curing beef and things like that. One critic's judgement was that Upton Sinclair aimed at the people's heart but hit them in the stomach. Sales of corned beef dropped after *The Jungle* came out. At the time of those writings, the twenties, there was another novelist, Sinclair Lewis. He was much more popular than Upton Sinclair, and he wrote some famous novels, or novels that are remembered. I can't remember any of them, but he was a more popular author.[16] His writings were more in line with the American Dream, the rather Republican view that you can make it if only you try hard enough. You too can be a millionaire. Upton Sinclair never wrote from that point of view at all. More read Sinclair Lewis than Upton Sinclair."

The position of nanny in Judge William Caldwell Coleman's household was a cut above the domestic jobs that had so far come her way in the States.

"A quite high-standing position. They hired her at some agency or other. A bit like a hiring fair: you went down to see who had got off the boat recently."[17]

The judge's wife, Elizabeth, took the lead in the interview, although the judge offered some searching questions, which is what he did for a living. Should Lou be accepted for the position, informed Mrs Coleman, the children in her charge would be William Jnr., a lad with a happy capacity for mischief, Robert, the clever one, Elizabeth, whose aloofness was a cloak for a delicate sensibility, and Susan, the pretty one. The latter grew up to be 'Mrs Richard E. Mooney' to the world at large.[18]

The successful candidate should have a good general education. When Judge Coleman asked Lou's school, she enunciated 'St. Margaret's' with such authority he let out a gasp. The name meant nothing, but its clarion iteration assured him that it was a very superior establishment. For her part, Mrs Coleman's impressions were favourable – she warmed

16 Sinclair Lewis has two entries in *1001 Books You Must Read Before You Die* (ed. Peter Boxall, Cassell, 2006) – *Babbitt* and *Main Street*. Upton Sinclair scores one, *The Jungle*: "A more socially important novel is hard to imagine" (p. 256).

17 Bill Leader, interview with the author, March 2, 2012. The source of all Bill's quotes until further notice.

18 The biographical information on the Colemans comes from conversations with Bill Leader, Jo Pye, Tom Leader and one useful written source: his obituary in the *Baltimore Sun*, January 13, 1968, cit. www.findagrave.com.

to Lou's amiable personality and her English accent.[19] But she needed to be reassured about the girl's politics.

"I want you to teach my children to hate the Germans," she said.

Lou was dismayed. Her internationalism came from her readings in the book room, and she knew that national boundaries would become irrelevant when class distinction was abolished. This probably wasn't the moment to break it to the Colemans.

"I can't do that," she said. "I don't hate Germans."[20]

The Colemans hired her despite this unfortunate lapse.

Judge Coleman didn't discriminate in his prejudices and was evenly tough on left and right. Union activists and the Ku Klux Klan were his pet peeves, although Germans edged ahead of both. On the one hand, he endorsed women's rights, particularly their right to vote, and fought for laws to protect public spaces. On the other (after Lou left his service), he struck down New Deal legislation and ordered Federal marshals to remove striking seamen members of the CIO[21] union from the S.S. Oakmar. His motto, repeated often and loudly, was that *national government should be supported by the people; people should not be supported by the national government.* In short, Judge William Coleman was die-hard Republican.

Coleman was nominated to the District Court for the District of Maryland by Calvin Coolidge on December 6, 1927. His court schedule was dominated by prohibition cases. Typically even-handed, he ruled that anti-alcohol enforcers *could* trace pipes to beer kegs situated outside the house named in a search warrant but *could not* raid premises solely on the evidence of their noses. The day after

19 "I don't know why she didn't have a Cockney accent. My dad had more of a London-y accent. It has been said that women don't have such a strong accent as the males in the family. In Liverpool it's very noticeable. The female Liverpool accent is quite different from the male Liverpool accent. Vowels are pronounced differently." Bill Leader, interview with the author, April 10, 2013.

20 This, like the street singer, seems to be a story vouchsafed only to Gloria. Gloria Dallas, interview with the author, February 27, 2013.

21 The Congress of Industrial Organizations, a federation of unions, organised industrial workers in the USA and Canada, and operated from 1935 to 1955. The Almanac Singers – Pete Seeger, Millard Lampell, Lee Hays and occasionally Woody Guthrie – gave them a song to spur the movement: *We'll fight for Harry Bridges and build the CIO…* ('The Ballad of Harry Bridges' by Hays, Lampell and Seeger).

Pearl Harbour he told a roomful of Germans and Italians (sobbing, we're told) that he could not formalise their naturalisation because of the imminent war between their countries and his. He refused to naturalise anybody who had left Germany after 1933, the year Hitler came to power. The High Court overturned the ruling.

As a patrician of the old school, Judge Coleman felt a duty of care towards his servants. He and his wife liked Lou. Mrs Coleman maintained the link and wrote her after she had left their employ. She told her about the children – one son became a pilot and the other became a pastor. Lou liked the Colemans and doted on the children.

"She had a very happy time for a few years there, in this young family. They went on holiday to the place where rich people go, Fishers Island, which is on the east coast above New York,[22] at that time a millionaire's playground. She got a sniff of that sort of thing. And she also got a sniff of the fact that the life of a federal judge was not all roses. Do you know the story of Sacco and Vanzetti?"

Oh Sacco Sacco
Oh Nicola Sacco
Oh Sacco Sacco
I just want to sing your name
Sacco Sacco Sacco Sacco Sacco,
Oh Sacco Nicola Sacco Sacco
I just want to sing your name

'I Just Want to Sing Your Name' by Woody Guthrie comes from *Ballads of Sacco and Vanzetti* on Folkways FH5485. The songs on the album were composed some two decades after the events described. Woody here croons the name Nicola Sacco as an incantation, a charm to ward off evil. Bill has commented on the simplicity of Woody's songs: "Guthrie could take a word or two and, by repeating them, create a song. The sort of song that looks ridiculous when written. *Tickle, tickle, tickle, tickle, jiggle, jiggle, jiggle, jiggle… O Sacco, Sacco, Nicola Sacco* but makes sense when sung."[23]

---

22 "Her summers, for instance, were spent, with the family, on up-market Fishers Island, a 9 miles by 1 mile strip of prosperously populated land nestling in Long Island Sound." From *If I Remember Writely*, a fragment of autobiography by Bill Leader.

23 Bill Leader, sleeve-notes to *Men at Work: Topic Sampler No. 3* (TPS 166, 1966), discussing 'The 1913 Massacre' by Ramblin' Jack Elliott.

Louisa with William Coleman Jnr at Fishers Island, 1927

Louisa as a young woman (collection: Jo Pye)

According to Lou, Judge Coleman was in contention for Judge Thayer's position. *Us varmints has got to get together too / Before Judge Thayer kills me and you*, sings Woody Guthrie, arguing for organised labour as a safeguard against the judiciary.

"My mother's employer was unknowingly on the shortlist of judges for the job," Bill explains. "It had to be a federal judge because murder was a federal rap. The first trial had caved in for some reason or other. And they thought, best try out of town. Mother remembers a message from a friend in a fairly high place saying to Judge Coleman to make himself scarce and distinctly unavailable. They were looking for some sucker to run this trial, and the sucker who ran the trial put himself at the risk of a bomb through the letterbox. Some anarchists were not quite as mild and meek as Sacco and Vanzetti."

It's a good story, but is it true? Judge Thayer presided at Sacco and Vanzetti's trail for the robbery and murders at South Braintree on September 11, 1920. The trial, which took place in Dedham courthouse, Massachusetts, began on May 31, 1921. The prosecution linked Braintree to an earlier robbery attempt (on December 24, 1919) at a shoe company in Bridgewater in which a gun was fired. Judge Thayer sentenced Vanzetti for twelve to fifteen years for Bridgewater, which was harsh in the extreme. Sacco was also in the frame for Bridgewater but was able to establish an alibi. He and Vanzetti were indicted for the robbery and murders at South Braintree and pronounced guilty on July 14, 1921.

Attempts by Sacco and Vanzetti's defence team to stage a new trial failed. They based their appeal on the prejudice of the judge, but, according to the Massachusetts' legal system, retrials needed the agreement of the original judge. Naturally, Judge Thayer demurred from publicly admitting he was bigoted, corrupt and racist, and denied a retrial. The miscarriage was plain: discredited evidence, the blatant unreliability of the witnesses, even the confession of the real murderer couldn't prevail against the implacability of Judge Thayer. Feelings ran high. When a last-minute appeal to the Supreme Court failed, a juror's home was bombed. The house was destroyed but the family were safe, reported the *Boston Traveler*.

Sacco received a higher voltage than Vanzetti when he came to be executed at Charlestown State Prison on August 23, 1927. The salt and water levels in his body were depleted because of a hunger strike, and salt and water make good conductors for electricity.

Judge William Coleman, a patrician of the old school, with Caspar the Cocker

The grown-up Coleman children, with Nancy, William Jnr's wife – l. to r. Betty, Nancy, William Jnr, Susan, Robert

The Colemans at Christmas – l. to r. Susan, Judge William Coleman, William Coleman Jnr, Elizabeth, William Coleman III, Nancy, Betty (collection: Bill Leader)

William Caldwell Coleman, however, was only appointed as a federal judge in April 1927, which exonerates him altogether. Significantly, another of Lou's good reads was *Boston*, a two-volume 'documentary novel' based on the Sacco and Vanzetti case by Upton Sinclair. *Boston* was published in 1928. So when exactly did Lou work for the Colemans? Jo is vague on the point.

"All I know about the Colemans is what she told me when I was a child. I have no idea how she ended up in Baltimore. Sorry."[24]

Back in her studio, poring through the family photographs, Jo draws my attention to a faded sepia print of a smiling woman in a twenties bathing costume, with a child atop her shoulders.

"That's definitely Aunt Lou, and that may be one of the Coleman children because no way is it William, and it's not one of her siblings either. I'm not 100% sure because I never knew them. They kept in touch with her for a long time. I remember when she went to live in Camden Town, and she didn't move there until 1960…"

Lou received Christmas cards from the Colemans well into the Camden Town period. Yet greeting cards are inappropriate to convey certain kinds of news.

Lieutenant William Caldwell Coleman Jnr was killed in a plane collision on a training exercise off the Florida coast in January 1945, aged 26. His brother, the Rev. Robert Henry Coleman, died aged 37 in 1957, "at his parents' home in Eccleston, Maryland". *The Baltimore Sun*, my source, doesn't elaborate, nor is further mention made of Elizabeth Coleman or Mrs Richard E. Mooney. Their father, upon retirement from the judiciary, continued in public life and served as a Republican member of the Baltimore District City Council until his death in January 1968. Elizabeth Brooke Coleman died in September that year, which is when the letters and cards stopped.

"That's Aunt Lou as a very young woman," resumes Jo, working her way through the black and white snapshots. "You can see the teeth. And she had hair like wire wool! Nothing would stay in it. When I was a child, she used to have these sorts of celluloid hairpins, like tortoiseshell, only they weren't. She'd stick them in, and they'd break all over the place, and it was my job to go round and pick them up. She had hairpins everywhere. Eventually I think my mother did persuade her to have it cut. But she used to rub it in with olive oil

24 Jo Pye, email to the author, October 21, 2016.

before she washed it, and then rub it in again, because if she didn't, it stood out in all directions. They all had hair like wire wool. But my dad's looked quite nice because it was auburn. That's a really good picture of her."

Lou is well represented in the photographs. The inscriptions on their backs reveal the documentarist: 'Aunt Lou in a wood', 'Aunt Lou, USA wood', and 'Aunt Lou in the USA.'

"Very helpful," says Jo, in a self-mocking tone.

They come from a large collection Cathy Davis inherited from her father, Harry. Jo helped identify some of the subjects when she visited the States in 2011, and Cathy, Nance's granddaughter, donated them to the official family archive in Sheffield.

"That may well have been when she was working for the Judge," says Jo, referring to the bather. "It looks the right sort of period."

Bill Snr seldom appears in the snaps.

The cousins always intended to marry. That was the understanding before Lou left for the States. It was natural that Bill Snr should want to join her. If Lou was radical in theory, Bill Snr was radical in practice, and a constant bugbear to the bosses. Yet Bill Snr is the more elusive figure, simply because Lou outlived him by so many years.

"He was a trade union shop steward in the Metal Box Company," says Gloria. "Very politically aware. If you asked him a question, he would answer you with a question, so you could work it out for yourself."[25]

"My father was fairly aggressively militant. Not over-aggressively militant. He was a trade unionist and he thought people who didn't show solidarity on the workshop floor should be treated with contempt… He came from south west London, Battersea way. One of a vast family."

Gloria is confident about Bill Snr's ethnicity.

"He was an Irishman really. He was an Irish descendant. He'd been to a meeting on one occasion, and he was on the top of a double-decker bus, and he was talking with a friend, and this bloke behind chipped in with something, and Big Bill said he thought he was going to throw him down the bus stairs for joining in like that. He had a temper."

I murmur surprise. He gave me the impression of a mild-mannered fellow.

---

25 Gloria Dallas, interview with the author, August 20, 2016. "Gloria worshipped Uncle Bill. She really did. Because he was a lovely man," says Jo.

"He was mild to most people, but if he didn't like somebody and they hadn't got good manners... He was going to throw him down the bus steps. 'What business have you got interrupting me and joining in my conversation? We weren't talking with you!' It was obviously someone he didn't like. I liked him very much."[26]

Bill is wary about the bus anecdote.

"He could ride on buses without wanting to throw people down the stairs. He was able to control himself to that extent. I mean that was an occasion, which I think was so exceptional that it gets repeated and then starts becoming something he did on a Friday night."[27]

"I think his original job was on the underground trains," hazards Gloria.

The 1911 census has Bill Snr as a motor examiner (according to the same source, his brother Jim worked on the 'electric railway'). The immigration form to the States gives his occupation as an engineer.

"He was an apprentice-trained mechanic of some sort. Eventually he became a skilled machine toolmaker and he worked in engineering factories all his life. That's what he did in America," says Gloria.

Bill Snr and Lou were reunited immediately upon Bill's arrival in America. What the lovers discussed when they met is not known, but shortly afterwards Lou resigned her post at the Colemans and quit Boston for New York. The couple married on March 24, 1928. Jo produces a copy of their marriage certificate, which has 'King's County, Clerks' Office, Brooklyn' in baroque writing on a very ornate letterhead. It reveals, intriguingly, that Bill Snr and Lou lived on the same street but not in the same house. The witnesses were Nance and her husband Harry.

The Irish community in New York was much like the Irish community in London only more so. Because cultural survival lay in avoiding the American melting-pot, the Irish of the diaspora tended to take great pride in their 'cursed nation' status and zealously observed old customs whilst inventing new ones, like an annual St Patrick's Day Parade. The cause of a united Ireland was sacred, and they were passionate about the 'auld music'. Musicians like Paddy Killoran and Michael Coleman made a far better living in the States

26 Gloria Dallas, interview with the author, August 20, 2016: also the source of the "fat lady" quote below.

27 Bill Leader, interview with the author, June 15, 2018.

New Jersey, 1930/31
Louisa and Bill

Bill Snr and Bill
(collection: Jo Pye)

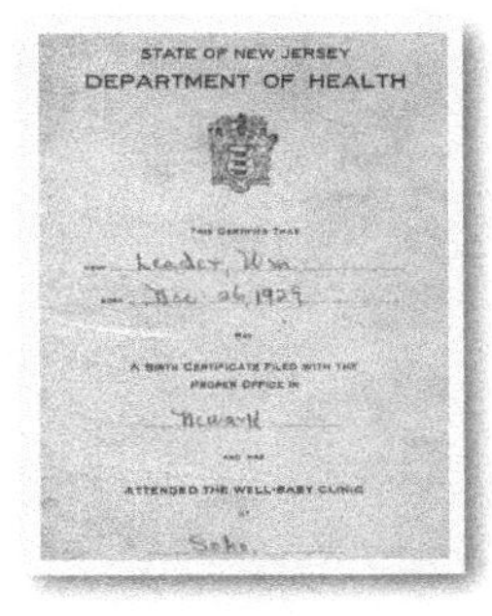

STATE OF NEW JERSEY
DEPARTMENT OF HEALTH

THIS CERTIFIES THAT

Leader, Wm

Dec 26, 1929

HAS

A BIRTH CERTIFICATE FILED WITH THE
PROPER OFFICE IN

Newark

AND HAS

ATTENDED THE WELL-BABY CLINIC

AT

Soho

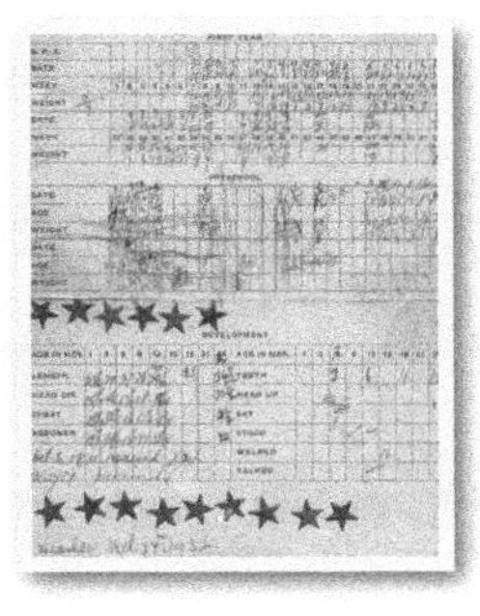

Bill's world
(collection: Bill Leader)

than ever they did in Ireland. Radio networks targeted shows at an Irish audience.

If Bill Snr wondered about this super patriotism, he kept it to himself. Gloria, who counts blue eyes, tapered fingers, a fondness for cards and suppressed temper as identifying signs of the true-born Irish, knows more about their home life in New Jersey than anyone else, Bill included, because of those long conversations with Lou in Shipley. She knew about the couple who would call to make a foursome at bridge, and about how the woman's minuscule appetite belied her obesity. One night Lou chanced to wander into the kitchen where the woman was washing-up – she was a cheery and kindly woman and always volunteered to do the washing-up – and found her ravenously devouring the leftovers. *Pleasure implies hunger, but hunger implies hope*, as the man said.[28]

So how Irish was Bill's dad?

"He was born in London. He spoke with a Cockney accent. All his siblings – he had many, many brothers, and two sisters – were born in London. They were nominally Catholics, but I don't think they were assiduous observers of the faith. So they were as Irish as you could be, being in a community of mainly Irish people, who stemmed from mainly Irish people. He was aware of his Irish background. He knew of things Irish. Customs and politics. Economic circumstances. But he was, what? One generation away from the bog. My mum's grandmother had this phrase, 'As Irish as the pigs in Connemara'. That was her definition of someone who was pretty bloody Irish.

"He wasn't a nationalist in the sense of being either English or Irish or anything else. He was a member of the working class. That's where he stood."[29]

The couple married and settled down in New Jersey where Bill was born on the Feast of Stephen, a day consecrated to the first Christian martyr. That is, December 26, Boxing Day, 1929.

> It was on the Feast of Stephen: the year was 1929. I have been told that it was snowing at the time, but whether it was deep or even crisp I don't know, and now it's too late to ask. I do, however, have documentary proof of my nativity

28 Louis MacNeice, 'Autumn Journal'.

29 Bill Leader, interview with the author, June 15, 2018.

> in the form of an ornate, gold sealed record of birth that announces "a certificate of birth has been filed with the City Clerk of the City of Newark, New Jersey."
>
> There was no stationary star over Newark, New Jersey that Christmas, but there had been a very dark cloud over Wall Street just two months previously, and the echo of the thunderous crash resonated around my cradle. It was the knock-on, shock-on effect of the Wall Street crash that drove my family back to England, to their natal London, to the run-down streets that they had determinedly put behind them just a few years previously.[30]

The decision to return was influenced by Bill Snr's activities as a trade union activist. Was he a Wobbly then?[31]

"No, because they tended to be a different sect of the left from communists. They were more what's called Trotskyite. I'm not quite sure of the subtle shadings of red, but, no, he wasn't a Wobbly."[32]

He was homesick for London and for London mild, as opposed to the stuff he illegally brewed in his bathtub. Lou, less of a beer fan, was sustained by pleasant memories of Fishers Island. Whereas Bill Snr had to contend with daily harassment, constantly coming up against the brick wall of management as he moved from factory to factory and from job to job, reflexively agitating for better conditions. With the Depression (it struck the month before Bill was born: "I didn't have anything to do with it," he pleads),[33] Bill Snr's hours increased and his wages fell, but he was employed; the only one in the street with a job. As a skilled mechanic, Bill Snr had no difficulty finding work. He was also in full-time guilt and alienation.

> My father was involved in politics. He had been a member of the Communist Party of Great Britain from its early days in the 1920s. He was a militant socialist and he had great faith in the future of the still young Union of Soviet Socialist Republics. He was aware of that country's need for skilled workers and, when the slump struck the US, he had seriously considered migrating to the USSR to aid

30 *If I Remember Writely*, ibid.

31 The Industrial Workers of the World (IWW), commonly known as 'Wobblies'.

32 Bill Leader, interview with the author, August 8, 2016.

33 Bill Leader, interview with the author, March 2, 2012.

> Stalin's new and exciting five-year plan. I would have had a different story to tell if he'd headed east that extra 1,000 miles to Moscow or been sent the 3,000 miles to Siberia. Instead we returned in 1931 to slump ridden England.[34]

So Bill Snr returned to England with his little family. It was 1931, and the Depression travelled with them. Bill Snr found a job working for Daimler in Coventry, but something went wrong: "I can't remember what," says Bill, a toddler at the time. After an interval in Coventry – "We lived in Meridan, the middle of the country, hence its name" – the Leaders headed to East London, where they had family.

"They didn't have any experience of anywhere else apart from good old London."[35]

Bill Snr, however, found it hard to adjust. Nothing in the States prepared him for conditions at Ford Dagenham, where, in emulation of work practices adopted by its parent in Detroit, goons paraded on a gantry above the factory floor, ready to enforce the required level of production. Machines always had to be manned, and workers were unable to leave their posts except by permission. "You can't stop making cars to shit," points out Bill, getting to the nub of it.

Ford Dagenham didn't have a strike-breaking militia like the National Guard at its disposal, as the parent company had, but the management did their best to dismantle the conditions and restraints painfully won in tortuous rounds of negotiations, and union officials were demoralised as a matter of course. Bill Snr stuck it out for four years.

Molly had crossed the ocean by this time.

"When either one of them came home on a brief visit, all was forgotten and forgiven, but my mother, who was not an uncharitable person, was never really charitable towards her two sisters in America. She got on perfectly well with the two sisters who never left the country."[36]

Lil was in service in the big house at Irstead Shoals, in Norfolk. The settlement was more like a row of four houses than a village, so

---

34 *If I Remember Writely*, ibid.

35 Bill Leader, interview with the author, March 2, 2012. For this, 'Meridan' and the 'shit' quote below.

36 Bill Leader, interview with the author, April 10, 2013. His contribution to the run-down on Lou's siblings, which follows interspersed by comments from Jo, comes from the same source.

the big house tended to dominate. She returned to London to have an illegitimate child, a boy, who was taken from her. George, a stoker in the Royal Navy from Irstead, married Lil in September 1930. Bill associates George and Lil with Horning, "at that time a rather quiet and remote place, and now a rich resort, an attractive little village on the Norfolk Broads. The house they moved into, and subsequently vacated, is one of the more expensive houses in the village."

Lil was evacuated to Horning during the war, where she was joined by her mother, Louisa Martha. It was a temporary arrangement, but Louisa Martha stayed on. As someone who had never left off toiling, she was ideal to look after Jim Jones' children. Jo gives me an account of Jim Jones…

As a native Londoner adrift in a regiment from Yorkshire – he was married to a woman from Gargrave, North Yorkshire, if that explains it – Jim Jones was hardened to the Yorkshire accent but found himself unequal to the challenge of the Norfolk dialect, billeted deep in 'hev yew gotta loight boy' country. In fact, the regiment occupied the field at the side of Lil and Louisa Martha's cottage in Horning. Jim actively sought their company for the reassuring lilt of Cockney vowels. Then Jim's wife left, leaving him with three small children. When Lil and George with daughter Lillian moved back to London, Louisa Martha stayed behind to take care of Jim's children, and, indeed, had legal custody of them. "I don't know how all that came about," admits Jo.

Meanwhile Kit, Lou's youngest sister, struggled to be taken seriously.

"She used to travel from Canning Town to Mornington Crescent, which is a bit of a journey, to a place known as The Black Cat, which is now called Greater London House. It was a bloody great place, with a mock-Egyptian front, owned by Carreras, the tobacco people. They made Craven A, the cigarettes that were good for your throat (believe me, I've seen the adverts). She travelled all that way to immerse herself in tobacco in order to get some strange circulatory illness."

To make a hard story easy, Kit moved to Horning to recuperate and married Jim Jones on October 25, 1947.

The East Anglia connection was reinforced when Bill married Lynne, who was born in Lowestoft and whose family still live in Norwich, but all that is much more recent.

"If we're going through my mum's siblings, which we seem to be doing, her brother Pat, Jo's father, was always in. She was never critical of him at all."

Pat went from the Merchant Navy to engineering. He worked on RAF stations and offshore defences installing heating and ventilation. Often, he would install a system, and the Luftwaffe would come and bomb the bulding, or strafe it with machine-gun fire, and Pat would have to start all over again. It was a sheer waste of time, as Sisyphean labours always are, but steady employment.

Then there was Harry, the boy who died in 1908, just turned eight.

"He looked like my father and there was only three years between them," says Jo. "I've always wondered whether that's why my dad was a little bit more distanced from the rest of the family. The brother that was closest to him died when Daddy was four. He couldn't have much memory of him, but he talked about him a lot. And he talked about Molly. She looked like him as well, and they were temperamentally very much alike. There were those three, Molly, Harry and Dad, and those three were ginger as well."

Molly was a rare visitor.

"Aunt Molly only came home once. It must have been 1960, something like that. Because a couple of her younger sisters had died, and she decided she better come home while they were still a few relatives left."

"That left two brothers," resumes Bill. "There was Jim, who was the second youngest, who spent his war out in Malta, a very nasty place to be during the war. I didn't see much of him as I was growing up, because he was in the Regular Army before the war, when it was easier to join the army than find a job."

Nicknamed 'Tubby', Jo remembers Jim "went out to Malta weighing about twenty stone and came back weighing about eight. They starved on Malta. He's one of the few that talked about his time in the army. He was an army staff driver. He was in Malta. He was in Italy. He says that he was at Monte Cassino, and I have no reason to disbelieve him. Except when I checked, it doesn't say that his regiment was there, but when I talked to a military historian friend, he said it doesn't mean a thing, because if you were short of people in one regiment you drafted them in from the ones that had got people."

Jim lived in Plumstead. In fact, he lived next door to where Tom, Bill's son, lives now (I'll be sure to tell you how this happened when I have the facts straightened in my head).

Ted and Louisa Martha
(Collection: Jo Pye)

"There was a bus, the 53, that went from Camden Town to Plumstead," says Bill. "You got on, and you could go all the way to Plumstead, which my mother used to do fairly frequently, partly for social reasons and partly because Jim's wife was a bit of a needlewoman. My mother, who was always working away at some sewing job or other, used to go for the day, and do it there.

"And that only leaves one other of mother's siblings, Ted, who she didn't get on with. We didn't see a lot of Ted, and nothing Ted said or did gained approval. Whereas Jim was a joker and an irresponsible sort of a lad, you didn't feel uncomfortable in his company. In fact, you felt welcome, because he had a ready line in patter and reminiscence. But Ted was more withdrawn, and less socially able, and didn't have much to commend him."

Ted was a long-distance lorry-driver and drove huge articulated trucks. He was spared wartime service because transport driving was a reserved occupation. He was bad tempered and had either a persecution mania or a superiority complex, Jo isn't sure which. The symptoms were indistinguishable.

"Thinking about it [says Jo], I can never decide whether Uncle Ted was paranoid or whether he just thought he was so great that everybody should do what he wanted, and then got in a huff when they didn't. And everybody thought that Aunt Violet was daft, but Aunt Violet wasn't daft. She was very good at managing him. If he did something… She criticised his painting of the stairs. She said he'd missed a bit or something. And he chucked the paint and brush. Aunt Violet went and made a cup of tea, ignored him, continued to ignore it, for two or three weeks, until he finally got around to clearing it up himself. Because she said, 'I wasn't going to do it!' Now that's not a silly woman. That's a clever woman!"

John, always known as Jack, was Lou's cousin and Bill Snr's brother. His father's side, Bill claims, "did more singing than my mother's side." Pressed to name some typical songs, Bill mentions 'Boiled Beef and Carrots' and 'Any Old Iron'.[37]

Jo's no-nonsense tone becomes hushed when she speaks of Uncle Jack. She has a year of birth, 1884, but no year of death, which worries her. His four children also lack final dates ('Tommy d. bet. 1948-1949' is the best she can do). Her voice becomes soft, as if someone has just woken her gently and asked her to recall her dream.

"Uncle Jack was around when I was a kid," she says. His wife, Hetty, was scary. Her nose ran to a sharp point, like a witch's.

"I don't know when he died. I know he had two sons, one of whom died quite young. I have a feeling… Uncle Bill and Aunt Lou came down and stayed with us in Hoddesdon sometime in the fifties. They came down for a funeral, and I remember being taken to the house. We didn't go to the funeral. It might have been Uncle Jack's funeral. My dad and Uncle George took me down to the railway. Down the bottom of the road from where they lived, there was this huge brick wall, and over the wall there was the railway line. I suppose I was about nine, ten. They took me down there to look at the trains because it was all so gloomy in the house.

"I don't know if that was Uncle Jack's funeral or whether that was his son Tommy, because Tommy died in 1948, 1949. He had chronic TB. He had TB and his wife had TB, and he'd been in the army in the war, and was said to have been imprisoned as a conscientious objector for a while, and then he somehow ended up in the army. But according to my mum, he went missing and when he turned up, he'd no idea where he'd been. No memory of anything. But Mum told me that when Tommy came home, wherever he had been, he had been well looked after. Because she said his clothes were washed and freshly pressed, and he'd obviously had a wash and a shave and been well fed. He hadn't

37 Bill Leader, interview with the author, April 10, 2013.

ended up in a ditch or anything. But he'd lost his memory, and nobody ever really knew what happened to him.[38]

"I met him briefly, when I was about four. It would be '47 or something like that. The first time I remember going to stay with Aunt Lou I'd been very ill. You name it, I'd had it in the previous six months, from scarlet fever to pneumonia and everything in-between. So we went to stay at Aunt Lou's, and for the first weeks that we were there, Tommy and his wife Grace, who also had chronic tuberculosis, were there. We didn't know, and they didn't know themselves, at the time. So we spent a week on holiday with them and then we went and visited them at Aunt Hetty's house a couple of times, and then not long after that, he died. But what I remember of him is a nice man."

Jo sighs and rouses herself from reverie.

"Do you want to go and have some lunch now, and we'll come back and see what's what..."

38 "My perhaps incorrect memory is that he reached the rank of Warrant Officer, which doesn't fit with going AWOL." Bill Leader, annotation, August 24, 2018.

Bill with a great aunt, two aunts, an uncle and a couple of cousins.
Back row – Bridget (Lou's paternal aunt), Tom and Julia (Bill Snr's brother and wife), Danny and Tommy (Tom and Julia's sons). Bill is in front and Kit is peering out of the window.
(collection: Jo Pye)

The full flood of 'England' swept him on from thought to thought. He felt the triumphant helplessness of a lover. Grey, uneven little fields, and small, ancient hedges rushed before him, wild flowers, elms and beeches, gentleness, sedate houses of red brick, proudly unassuming, a countryside of rambling hills and friendly copses. He seemed to be raised high looking down on a landscape compounded by the western view from the Cotwolds, and the Weald, and the high land in Wiltshire, and the Midlands seen from the hills above Princes Risborough. And all this to the accompaniment of tunes heard long ago, an intolerable number of them being hymns. There was, in his mind, a confused multitude of faces, to most of which he could not put a name.

RUPERT BROOKE, 'THOUGHTS OF HOME'

III

Ballyhoo and Trauma
(Do You Want Us to Lose the War)

"It was a very strange period, the thirties. All sorts of things were going on. Mussolini was messing about in Ethiopia, Franco kicked off in Spain…"[1]

There goes Bill, understating again. By 1934 Hitler had assumed absolute power and German rearmament was an open secret. In 1935 Italy invaded Ethiopia, one of two African states not colonised by the west, without bothering to declare war. In 1936, 22,000 troops marched into the Rhineland on Hitler's orders. Japan invaded China in the same year. The mid-decade saw the first trials of the Old Bolsheviks in Moscow. Stalin could no longer tell his friends from his enemies, so he resolved to kill them all. And in July 1936 the Spanish Army, augmented by Moroccan mercenaries and abetted by the Nazis, launched a coup against the re-elected Republican government on a tsunami of blood.

1 Bill Leader, interview with the author, March 2, 2012.

The UK had the Means Test, Hunger Marchers, Mosley rallies and a great deal of royal rigmarole.

"King George V died in 1935," says Bill and instantly corrects himself. "Wait a minute, I'm getting this wrong. King George V had his Silver Jubilee in 1935. He didn't die. Far from it. He had a big celebration. There were stamps, twice as long as normal stamps, and there were mugs. There were, of course, street parties, which is how the British celebrate these things. I've never been to a street party, but I've read about them and seen them on newsreels, so I know that that's what we do. We had street parties for George V's Silver Jubilee."

To avoid street parties for four score years and ten shows integrity, forward planning and admirable anti-sociability. Bill's upbringing served him well in this respect. In an age when baptism followed birth as routinely as night followed day, Bill remained defiantly unchristian. He was the classroom atheist and junior anti-royalist. If anything, his rebelliousness has grown with the years. Not that Bill is the kind of man to shove his opinions down your throat…

"George V died in 1936, the stupid sod! There was mourning, solemn music, and John Snagge put on a particularly sarcophagal voice when announcing. In 1936 there was a new king who went and chucked the job in: he abdicated. And we had a coronation. George VI came on. So in the middle thirties we had our nose rubbed in royalty. We were either celebrating the fact that he had been with us for so long, bemoaning the fact that he was no longer with us, giving the go to a new one who failed miserably, and then giving the go to another one who couldn't put two words together. It was a strange old period."

Bill's pronouncements on public affairs often have this quality: of indignation refined by irony, expressed in gentlemanly terms that might be a conscious embodiment of the ironical spirit.

"You had to believe in something, so you believed in the British Empire. It was a fine thing to believe in."[2]

It was an age of deference, when a boy soprano could flourish, but not for long. The knowledge that a sweet voice is a fleeting glory adds a dash of what the Japanese call *mono no aware* ("the sadness inherent in things") to the art of the boy soprano. HMV were doing very nicely out of Master Ernest Lough, who, in 1926, provided the company with a runaway hit in 'O For the Wings of a Dove'. Sadly, Master Ernest

2 Bill Leader, interview with the author, June 8, 2012.

Lough's season in the sun had passed. The hunt was on to find his successor. HMV's talent scouts were alerted to the existence of Master Ray Kinsey when a silver cup at a talent contest in Blackpool flushed him into the open.

"Ray Kinsey had been a boy soprano," says Bill, who knew him well in later life. "I don't know quite how old he would have been, but young enough to be a boy soprano."

When Kinsey managed Livingston Studio in Barnet, and Bill was a busy client, Bill transferred Master Ray's original 78 discs onto tape as a favour to his friend. They came from Ray's thirteenth year, when he was at the height of his powers.

The first of these discs coupled two Handel arias, 'Rejoice Greatly' and 'Let the Bright Seraphim', and was recorded in January 1933 at Kingsway Hall, a Methodist church in Holborn reputed to have the best acoustics in London. The prodigy's technique still impresses. His crystal-clear enunciation never falters on the rising thirds and falling semi-tones of the challenging coloratura passage of 'Rejoice Greatly'. This was piety of the gutsiest kind. Master Raymond, you felt, would beat Master Ernest black and blue in a schoolyard scrap.

'Ave Maria', from a later session, preserves a sound too pure to survive, and is an emblem of a world already irretrievably lost. Having recorded six sides for HMV in the annus mirabilis of 1933[3] – 'Rejoice Greatly' was released in January and the last, 'Oh, For a Closer Walk With God', came out in September – Kinsey's voice broke like an egg dropped onto a hard floor.

It was the year Lou Leader "…argued her way onto the London County Council housing list, getting us a newly brick-built terrace house, right on the edge of the LCC's slum clearance housing estate at Becontree, Essex, which was then, and I think still is, the largest public housing development in the world."

> Thousands of Ford's employees lived on the housing estate, and our house at Basedale Road was on one edge, near an open area we called Matchstick Island. To our back there was a seemingly endless tangle of streets, lined with well-built, brick-built terrace houses, all new and neat and spick

3 The original HMV 78 discs by Master Raymond Kinsey have been digitally restored and scattered across the six volumes of *The Better Land* (Amphion Recordings).

> and span. Sixty years on I revisited the place. It looked much the same: it had lasted well. The most noticeable difference was that the previously empty streets were now tightly lined with cars. The planners of the twenties had not anticipated the future need for garages, nor anticipated that the many Ford workers who lived there would have either bought their cars ex-factory at favourable rates or acquired them in kit form, enterprisingly smuggling them out of the factory gate, a part at a time, for extra mural assembly.[4]

"It was just us and then a green field, and then there was Barking."[5]

The Peace Council had a presence in Barking, and Lou was a committed supporter. We know how she channelled her political anger into shopping at the Co-op, selling and knitting woollen things for the *Daily Worker* and joining peace groups. Lou was a paragon of social awareness.

"I was three when I arrived there; I was seven when I left, so I was quite young. I went to school for the first time: things like that. But I was beginning to become aware of political organisations. As a little kid, I got dragged around everywhere my mum went to, so I was frequently dragged along to meetings or po-faced conferences and lectures that were held in some institution or other. I'd be taken along and told to sit there quietly, and then I would sneak out. I knew the positions of the emergency exits and toilets of every building in the area, from wandering around while my mother and father were at meetings."

The Peace Council, according to its constitution, was non-political and non-sectarian. A front for the Communist Party, then?

"It was driven, I think, by left-wing intellectuals who felt very strongly that things had to be done to try and save the world. And this was one of the areas where they did it. It was tremendously active. They did all sorts of things: campaigning, organising meetings, organising demonstrations, organising pageants and things like that."

The leading light of the Barking branch was Ariel Levin, a Cambridge graduate who opted for organising rather than spying.

"He was more of an organiser than an agitator. He was a brilliant organiser. He had a golden touch. He was able to establish contact with all sorts of people, and was around our house frequently, our little

4 *If I Remember Writely*, ibid. Also the source of the preceding quote.

5 Bill Leader, interview with the author, March 2, 2012.

council house on the edge of the Becontree estate. He talked to all sorts of people, in all sorts of walks of life around the locality. He was able to do that. Gifted. He went to Spain and got killed. I've never seen a reference to him in any political biographies or political commentaries. If he had lived, I think we would know his name."

Ariel Levin died on a battlefield in Aragón in March 1938.[6]

The Spanish Civil War was the great cause-célèbre of the left between the wars. Harry Pollitt, general secretary of the Communist Party of Great Britain, wrote the pamphlets *Save Spain From Fascism* and *Arms For Spain* to garner support.[7] In truth, the British left was desperately divided over the Spanish Civil War. According to which faction you belonged to, it was either a struggle for democracy (CPGB) or a revolutionary struggle (ILP/POUM). The argument was not merely theoretical. In the Spanish maelstrom, subtle shades of red became a matter of life and death. The schism degenerated to the point where the *Daily Worker* could describe ILP volunteers in Spain as a "stain on the honour of the British working class".[8]

Bill Snr and Lou were proud of having a relative who fought in the Spanish Civil War. He was Dick Baker, a nephew (the son of Bill Snr's sister, Anne). Dick survived, although he was wounded in the arm. Lou and Bill Snr did their bit by fostering a refugee from the Basque country, but not before the move to Mottingham.

> We were four years at Becontree, then dad went across the Thames, from Essex to Kent, to the Vickers Armstrong Company's Crayford works. We all moved, in 1937, to the slum clearance estate that the London County Council had

---

6 See *Help Spain: Voluntaries, Británicos e Irlandeses en la Guerra Civil Española* by Richard Baxell, Angela Jackson and Jim Jump, with an introduction by Ángel Viñas. Ariel Levin's fate is given in a list of fallen British volunteers.

7 These pamphlets were exhibited in a glass case providing context for the permanent display of 'Guernica' at the Reina Sofia in Madrid. They were displaced by a marvellous exhibition, *Pity and Terror: Picasso's Path to Guernica* (April-September 2017). By then I was so immersed in my subject that Picasso reminded me of Gloria Dallas, rather than the other way around. Picasso's 'Mandoline et guitare' and Gloria's *Brownie McGhee and Sonny Terry* (12T39) might be father and daughter.

8 *Britain and the Spanish Civil War* by Tom Buchanan (Cambridge University Press, 1997). See also George Orwell's *Homage to Catalonia* (1952; my copy is the 1962 Penguin edition), a vivid account from a POUM volunteer, and great literature.

Bill, Bill Snr and José Luis Tovar González outside 1 Calcott Walk in Mottingham (illustration: Peter Seal)

> built on the old Court Farm at Mottingham, on the London edge of Kent. Vickers was changing from car to armament manufacture at the time. And it was a good time to be in armaments. The world was in churn: Civil War in Spain; trouble in Palestine; Japan in China; Italy in Ethiopia and Germany was re-arming.[9]

José Luis Tovar González was a lad of Bill's age (10) from San Sebastián, placed with the Leaders by the Basque Children's Committee, part of the National Joint Committee for Spanish Relief, which co-ordinated aid to Spain. The evacuation of Basque Refugees – the children of Republican homes, not all of them Basque – commenced in 1937. Boats chartered by the Basque government embarked from Bilbao to Britain, Belgium, the Soviet Union and Mexico. José was one of nearly 4,000 children who arrived at Southampton Docks on May 23, 1937, on the cruise-ship *Habana*.[10] A nice little boy, according to Maggie. José was charming and handsome, confirms Bill, and quickly picked up the language. Yet Maggie detected an atmosphere.[11] She couldn't put her finger on the cause. I can reveal that it was Bill.

"We didn't get along, because in my immature way, I just treated him as a rival for my parents' affection. I didn't go and kick him when he wasn't looking or anything like that. It was just my overall, unsympathetic attitude towards him. Very infantile. But I *was* only an infant."[12]

José stayed with the Leaders for a matter of months in 1940, before the move to Keighley cut short the fostering and he was returned to the care of the Basque Children's Committee.

The story of the child evacuees of the Spanish Civil War is not as well-known as the Kindertransport, and not quite as satisfying. For one beast cornered in Berlin, another was allowed to slouch towards power in Madrid. Some of the 4,000 Basque Children of '37 were adopted and grew up in the UK. Some put down roots and settled; others tried to assimilate but failed and were alienated for life. Some more returned

---

9 *If I Remember Writely*, ibid.

10 Extrapolated from the wikipedia account – https://en.wikipedia.org/wiki/Evacuation_of_children_in_the_Spanish_Civil_War

11 Jo Pye, interview with the author, December 12, 2013.

12 Bill Leader, interview with the author, February 15, 2013.

and were just as alienated in Spain. José Tovar went back and departed from Bill's life as he now exits from these pages, a soul in transit. His thoughts about exile, England, and childhood associate Bill are not known but might reasonably be guessed.[13]

On October 6, 1936, Oswald Mosley and his Blackshirts were routed by the combined might of Jewish and Irish workers from the hitherto divided ethnic tribes of East London. A bookseller, Jack Firestein (whom we shall meet later), was involved. Artist Jim Boswell gave an eyewitness account of the act of resistance known as the Battle of Cable Street:

> Two policemen are wrestling with a young boy. His face is streaming with blood, and behind them an inspector jumps about, truncheon raised, trying to get in a blow. The boy is shouting, "Unity, unity, unity, unity." It is the word of the day. Under the blows of the police and the provocation of the Blackshirts the East End is being welded into a solid, united mass with a single idea – Mosley shall not pass. The police finally get their man away. It took about fifty of them to make one arrest there.[14]

Boswell's article appeared as a double page spread in *The Eye*, the house newspaper of left-wing publishers Lawrence and Wishart.

"Jim never fought anyone. He went to Cable Street armed with nothing but a sketch book and a fountain pen and was treated with respect by the beat policeman because he was wearing a clean white shirt and a tie. I can probably find a picture of him around 1936 but it will be benign and benevolent. Jim was universally described as genial," says Jim's daughter Sal.[15]

He had just started work as a graphic designer at Shell when Cable Street kicked off. The job "…became necessary because he had to earn a living now he had a baby".[16] That baby was Sal Shuel neé Boswell, born in 1936. Nor was the new job a capitulation to corporate

13 The Basque Children of '37 Association is an organisation dedicated to keeping the story of the 4,000 refugee children alive. See https://www.basquechildren.org for further information.

14 *The Eye*, no. 7 (Sept.-Nov. 1936). Cit. http://www.jboswell.org.uk/cable.php.

15 Sally Shuel, interview with the author, October 17, 2017.

16 Sally Shuel, email to the author, May 4, 2020.

"Two policemen are wrestling with a young boy..." The Battle of Cable Street, as witnessed and documented by Jim Boswell (picture courtesy of Sal Shuel)

Jim and Sal
(picture courtesy of Sall Shuel)

values. Shell was a successful and tolerant multinational guided in its recruitment by merit rather than politics. "It was accepted that the creative people had politics contrary to the managers. Sometimes the client was caught in the middle. One slogan got up by an ad agency for a politically innocent sales campaign read *ALL OUT MAY DAY*."[17] And though nowadays we might look askance at a campaign to get people touring Britain by car and refuelling with Shell petrol, it generated great posters and gave a wage to otherwise unemployable types, like artists. Material security enabled Jim to focus on the important thing – the fight against fascism. He formed the Artists' International Association in 1933 to organise exhibitions with this aim, and its original name – the International Organisation of Artists for Revolutionary Proletarian Art – nailed Boswell's colours to the mast.

The Left Review, under editor Randall Swingler (he wrote radical plays and translated Soviet songs into English, and was involved in the Topic Record Club), enlisted literary talents and cartoonists to the cause. Among the latter were James Boswell, James Fitton and James Holland, collectively known as 'the Three Jameses'. H.G. Wells, Bernard Shaw and J.B. Priestley contributed to the *Daily Worker*. The Left Book Club, a good idea by publisher Victor Gollancz, offered a monthly book choice for half a crown. These "...little orange-coloured, far-from-poetical publications" (© John Betjeman)[18] included *Days of Contempt* by André Malraux and *The Road to Wigan Pier* by George Orwell.

In fact, left-wing cultural groups were springing up everywhere, in response to the rolling crisis of the thirties. By the end of the decade the Left Book Club, the Workers' Music Association, Artists' International Association, the Birmingham Clarion Choir Singers and Unity Theatre were active in their respective fields – literature, music, visual art, communal singing and theatre. The existence of each was predicated on the belief that art could make a difference. Together, they turned the "very strange period" of the thirties into a Golden Age of Radicalism. Those involved knew each other and worked together, and some of the

17 'Boswell Remembered,: A Memoir by James Friell', cit. http://www.jboswell.org.uk/friell.php.

18 John Betjeman, 'Wartime Tastes in Reading', Home Service, broadcast September 4, 1944. Published in *Trains and Buttered Toast*, John Betjeman (John Murray, 2007), p. 147.

groups formally affiliated, resulting in creative interaction across the disciplines.

Alan Bush, later to be professor of composition at the Royal Academy of Music, co-founded the Workers' Music Association in 1936. He lived in Berlin in the early thirties, where he studied piano with Artur Schnabel, and adopted German models. Wife Nancy, like Swingler, translated Russian songs and texts into English; she wrote the libretti for Alan's operas. If you like, she was Bertolt Brecht to Alan's Kurt Weill. Historically, opera was *the* art-form to harness revolutionary force, and this was the tradition they aspired to emulate. When the Bushes' children's opera *The Press Gang or The Escap'd Apprentice* was staged at the Rudolf Steiner Hall in 1948,[19] Unity Theatre's designers provided the sets.

The invasion of the Soviet Union by Germany in June 1941, followed in short order by the Anglo-Soviet Treaty, galvanised the WMA. The first issue of *Vox Pop*, the WMA bulletin, in January 1942, contained a manifesto issued by the 'Executive Council', almost certainly written by WMA General Secretary Will Sahnow). Workers were enjoined to form factory and civil defence music groups, with the Red Army Choir as a model and inspiration. Music was effectively on a war footing. The propaganda value of topical song was embraced:

> A drive must be made for new songs dealing with each new situation that develops. It is bad enough to leave all the fighting to the Soviet people – don't let us leave the songwriting as well.[20]

(It succeeded, in so far as Plessey Aircraft workers banded together to write a theme song for the factory.)

In addition, members were urged to display and sell WMA music publications and records at social gatherings. *Popular Soviet Songs* sold 8,000 copies within months of publication in 1941.[21] It remains a source of controversy to this day:

> Sheet Music as Propaganda – 'Popular Soviet Songs – Words and Music' complete with hammer and sickle in RED! Pure

19 'Three Operas Produced by the WMA', flyer, 1948 (courtesy of the Political Song Collection, Glasgow University).

20 WMA Manifesto, *Vox Pop* no. 1, January 1942.

21 *Vox Pop* no. 1, January 1942. The same issue made mention of Alan Bush's leaving-do at the Bonnington Hotel, below.

> Stalinist Russia product with unbelievable lyrics e.g. 'I know not of an-y other country / Where man's freedom can with ours com-pare' – *I won't even go on with this* – 'Our land so beau-ti-ful and free…' From 'The Land of Freedom' two-part song with piano accompaniment. Published in 1941 by the WORKERS' MUSIC ASSOCIATION of 9 Gt Newport Street London WC2 translated into English by various hands. A piece of history!! None of the seven songs could be called 'Popular' – a pack of lies would be more appropriate![22]

Yes, but the songbook was published only slightly in advance of the Anglo-Soviet Treaty of May 1942, when Britain and the Soviet Union joined against a common enemy. *Popular Soviet Songs* was the WMA's heartfelt contribution to the war effort. Nor could the commitment of the WMA's founder to the anti-fascist struggle be doubted. In November 1941, a farewell party was held at the Bonnington Hotel, Bloomsbury, on the eve of Alan Bush's departure for the armed forces. Will Sahnow gave a speech about losing "a guide, philosopher and friend", but pledged to carry on the work on the home front.

Bill is about to drop a bombshell remark:

"More good things came out of Unity Theatre than ever came out of the Workers' Music Association."[23]

The statement is the more devastating for Bill's casual manner.

"I like to think I was there the day Unity opened," he continues. "I went to the Working Class Movement Library [at the Crescent, Salford], to check that out, and apparently it wasn't the day it opened. But that's why we were there. We were living in Mottingham. It was in 1937. I would be seven. There was a celebratory meeting and a performance of a Soviet play called *Aristocrats*, which is about building dams." And, indulging his weakness for puns: "A dam fine work!"

In fact *Aristocrats* was the first work staged at Unity's new venue at 150 Chalton Street, just off Goldington Crescent, Somers Town, a step-up from the church hall in nearby Britannia Street, where the workers' theatre had opened the previous year (1936), and which was

22 The sales pitch for a copy of *Popular Soviet Songs* offered for sale on eBay in 2015. The author managed to get it at a knockdown price from the seller. He was happy to let it go.

23 Bill Leader, interview with the author, November 28, 2014.

too small to contain Paul Robeson (he starred in *Plant in the Sun* with Alfie Bass). Somers Town, a neighbourhood that had gained in boisterousness what it had lost in gentility, was a good locale for a people's theatre.[24]

Aristocrats was a play by Nikolai Pogodin on an epic scale, a didactic piece of theatre that aimed to demonstrate how the dignity of labour – specifically the construction of the White Sea Canal – was capable of turning unruly prisoners into model citizens. There's no gainsaying Russia's grip on the popular imagination in 1937. Unity Theatre itself was modelled on Nikolay Okhlopkov's Realistic Theatre in Moscow, and *Aristocrats*' producer Herbert Marshall had worked with Okhlopkov and Meyerhold.[25]

"The only thing I remember was that finally the canal was opened and the water ran through, and they'd got a screen, and they'd got a blue textured thing moving down the screen, and that represented water. The audience clapped, and I couldn't figure out why they were clapping, because to me it looked nothing like water. They could have done better, I think."[26]

Seven-year-olds aren't as easy to deceive as a lot of grown-ups.

Unity people ran to deadlines, whereas Topic people, Bill explains, tended to take their time. And Unity turned working-class experience into drama that was fresh, visceral and often funny.

---

24 John Foreman, born near Euston Station, describes cramped conditions and a transient population. "There was a family in the basement, that's two rooms. We had three rooms. There was always somebody in the ground floor front. Everybody used the loo at the back. There was a couple on the top floor with a son, Mr and Mrs Whiteman, and people came and went. You paid your rent by the room. At the moment, the world is full of horrors. I can't remember any of that in my street when I was a kid. The main reason was, in our street you had very little privacy. There was so many people in one place. That's why every street had a public house. You needed a public house because you were so overloaded, you were so compressed. But the sheer weight of numbers made it much more difficult for unpleasant things to happen." John Foreman, interview with the author, April 20, 2016.

25 According to a letter from P. C. Hornby, artistic director of Unity, to *The Spectator* (September 17, 1937) anticipating the staging of *Aristocrats*. See also http://archive.spectator.co.uk/article/17th-september-1937/21/pogodins-aristocrats-. www.wcml.org.uk/our-collections/creativity-and-culture/drama-and-literature/london-unity-theatre/.

26 Bill Leader, interview with the author, May 5, 2016.

"They were quite well-known for their political revues, because it was the period of the intimate revue," says Bill.

The revues were polemical *and* rib-tickling and had pointed titles like *Swinging to the Left*.

"They also put on a pantomime. I think they put on several, but one called *Babes in the Wood* became quite famous. I've got a Decca 78 of two of the songs. One on each side, as is the custom. The male lead was a fellow called Bill Rowbotham, who grew up to be Bill Owen. I have a record of him before he was Bill Owen, singing a song called 'Love on the Dole'. It's a duet. Both songs are duets. The other side is called 'Affiliate With Me'. Another love song, you see."

(In the fullness of time this live wire turned into Compo, the adorable old codger from BBC sitcom *Last of the Summer Wine*.)

Babes in the Wood mocked Hitler and Mussolini, which was permitted, even encouraged, in the run-up to war. The trouble was, Unity subjected PM Neville Chamberlain to the same swingeing irreverence. The cradle of satire, it seems, was not Cambridge but Somers Town.

Then Bill Snr and Lou took Bill along to see *A Festival of Music for the People* at the Albert Hall, a spectacular funded by the London Co-operative Society, produced by André van Gyseghem, with artistic input from the Workers' Music Association.

Grand concerts in aid of political causes had become an annual fixture at the Albert Hall, but now, in 1939, with Spain lost, campaigning energy was diverted to get aid to China and the Soviet Union. Bill describes André van Gyseghem as 'the miracle man'.

"He was a very considerable theatrical producer, and one of his specialties was being able to fill large halls and deal with space in a very imaginative way. This was in the thirties."[27]

(Later Van Gyseghem would become Number Two, one of several, in the cult sixties TV series *The Prisoner*. The dialogue between Number Six and Number Two in the opening sequence, if you remember, was unvarying:

"Whose side are you on?"

"That would be telling.")

The WMA didn't initiate these events, but they knew which musicians were sympathetic to the cause (there were a fair few;

27 Bill Leader, interview with the author, December 1, 2017.

London Philharmonic was about to turn itself into a co-operative), and in Alan Bush they had an in-house composer. Bush had form in large-scale works, but *The Pageant of Labour* at Crystal Palace in 1934, and *A Pageant of Co-operation* at Wembley in 1938 were mere try-outs for *A Festival of Music for the People*. Paul Robeson, as Unity patrons could testify, could storm heaven's gates unaided, and here he was one among five hundred voices. The recruitment of an all-woman orchestra was an unlikely nod to feminism, but the world was not ready for a female conductor in 1939. The composer conducted his own uncompromisingly atonal score, after the manner of the Second Viennese School. The child Bill, with the same clear-eyed vision that saw through *Aristocrats*, was unimpressed.

"Bloody miles to the Albert Hall. The Albert Hall has not got any convenient tube stations anywhere near it. Even now, when they've been tunnelling under London like mad, they've not bothered to get to the Albert Hall yet."

Lou and Bill Snr, with young Bill, allowed themselves plenty of time.

"If you get to South Kensington, which is where we got to, you walk out down a long, long tunnel, which gives you access to several museums, including the V&A. We had a minute or two to spare so we popped in. I saw David standing there bollock naked. I was amazed. 'By gum!' I thought. We eventually got to the Royal Albert Hall, and we were in good time. They hadn't opened the doors, but there were hundreds of other people who were there in good time, and, to keep our spirits up, there was singing.

"So we all stood there, and we sang these things, like…

We'll make the king and queen clean out the public lavatory
We'll make the king and queen clean out the public lavatory
We'll make the king and queen clean out the public lavatory
When the Red Revolution comes.

That sort of thing. We sang 'Bless 'Em All', which became a wartime song. They'd changed the words to make it acceptable, and it was being played by dance bands and all sorts of people. Oh and the one that, if you were clever, you could extemporise forever – 'The Quartermaster's Store'. Have you ever stood in line outside the Albert Hall and sung 'The Quartermaster's Store'?

There was cheese, cheese, wafting on the breeze
In the stooore, in the stoooooore

There was cheese, cheese, wafting on the breeze
In the Quartermaster's Store.

My eyes are dim, I cannot see
I have not brought my specs with me
I haaave not broooouuughhht my-a specs with me

"Now all you have to do is find another commodity and find something derogatory about it that rhymes and you've got a whole other verse, and this extremely lengthy chorus, so you could spend bloody hours singing 'The Quartermaster's Store'. It's a beautiful song for standing in the cold outside the Albert Hall. I remember that."

Other songs commented on the hot topics of the day.

"Oh, there was a parody of 'The Policeman's Holiday', a novelty song, that was derogatory towards our then prime minister, the Right Honourable Neville Chamberlain. One of the attractive bits of 'The Policeman's Holiday' was the end of the chorus, which went:

Take your cards and beat it, Neville
GET OUT!

"Not only was it my first time in the Albert Hall, but we had access to the boxes – the sponsored boxes that belong only to the people who paid money when the Hall was built, and passed down in perpetuity to the first-born and so on. I remember we sat at one of these boxes in the Albert Hall. It was from there that I saw Paul Robeson, and Alan Bush waving his arms around, and the ladies' string orchestra. I'd never seen a ladies' string orchestra before. They were all sitting there, looking all prim and proper and pretty. Alan Bush at that time had a beard. I was surprised when I saw him later, and it was many years before I saw him again, that he didn't have a beard. I thought, 'Some people!' And I can't remember anything else really."

Paul Robeson, five hundred voices, a women's orchestra and Alan Bush's musical apocalypse washed over Bill's tousled head.

"I think they were playing something that was a little complicated for my nine-year-old me, not terribly educated in these matters. I don't remember specifically. I don't remember what Paul Robeson sang."

"At such an impressionable age too!" (This is my remark, and it strikes as priggish as I set it down.)

"I'm at an impressionable age now. If you want to impress me, do it now."[28]

Mottingham, where the family lived at 1 Calcott Walk, was ideal for outdoor-inclined boys.

"It was a London County Council housing estate, but very attractively positioned. If I came out of Calcott Walk – it was called a walk, not a road; it was a bunch of houses around a bit of green – and walked a few hundred yards to the right, I was in extensive woods. And if I turned left, and walked about 400 yards, I was in another lot of extensive woods. It was a very attractive place."[29]

Gloria says, admiringly, "He had a bike and he used to go all over. If his mother had known where he went!"[30]

This mix of suburbia and countryside recalls another William. The resemblance has been noted – Jo recently gave Bill a *William* book for Christmas. One scene, indelibly stamped on my memory, has William eating apples on the top of a double-decker bus and dropping the cores on people's heads. This from memory, so no reference, and it might not be word perfect: "He felt no exultation when the apple-core landed on top of someone's head, which is where he always aimed, just the impersonal satisfaction of the artist who knows his work is good."

William Brown did his bit for the war effort, and his own unfortunate flirtation with fascism has been forgiven (or suppressed: 'William and the Nasties' was omitted from reprints of *William – The Detective*, the 1935 entry in the series). Bill plays down any connection with the immortal schoolboy. "Oh yeah, I went through the William stuff," he says dispassionately, as if it was a long time ago, which it was.[31]

As wonderful as *William* is, delineating the horror of things creeping in shadows is beyond Richmal Crompton's scope.

---

28 Bill Leader, interview with the author, December 1, 2017.

29 Bill Leader, interview with the author, March 2, 2012.

30 Gloria Dallas, interview with the author, February 27, 2013. Albeit, she was likely thinking more of Yorkshire than Kent.

31 Bill Leader, interview with the author, April 10, 2013. And, of a slightly later period: "I went through the children's section of the Victoria Road Library, that Sir Titus Salt had endowed us with. I read all of Conan Doyle's short stories. I had to renew that a few times. And the funny stories of Edgar Allan Poe, as well as the frightening ones." It came as news to me that Edgar Allan Poe wrote funny stories.

Monteagle Infants School, Dagenham, c. 1935: Bill, almost in the middle, third row, standing, in a jacket, braces above and braces to the right (collection: Bill Leader)

Bill and Hercules

His first pair of long trousers

Bill in a boat in Shipley Glen
(collection: Bill Leader)

"God knows who or what was in the shadows about the house. There was a period when I was convinced it was IRA people in the shadows."

The IRA bombing campaign of 1939, soon eclipsed by war against Germany and nightly visits from the Luftwaffe, struck terror in Bill's boyish heart.

"That was after I'd managed to convince myself that it wasn't a burglar. Because I'd read this book, well, a story in a kiddie's annual. This kid was alone at home and a man knocked on the door and said, 'You've got a gas leak.' He said, 'I need to go down in your cellar and sort it out.' The boy was in alone. His parents were out. So he let the man in, and the man went down into the cellar, and the boy went up to his bedroom, where he had been, and was reading. Then he realised that he was locked in his bedroom. He realised that the man wasn't looking for gas leaks – he was looking for the jewels in the house. He was a burglar!

"But he was a very resourceful lad, and he went to his bedroom window, and taking the torch that he had, and using the well-known morse code, he flashed the message across to his friend whose bedroom window faced his, and managed to convey that there was a burglar in the house, and he was locked in his room. A policeman came around and sorted it all out.

"This scared the shit out of me. Because, ignoring the fact that we didn't have any jewels, I didn't know the morse code. What was I going to do? I thought, 'Gosh, how vulnerable we are when we don't know the morse code.' It didn't worry me enough to actually learn the morse code. I wasn't that sort of lad. It didn't even worry me enough to make sure I had a torch by my bed."[32]

William Brown, a working-class hero trapped in a middle-class family, had trouble with the authoritarian wing of the grown-ups – the teacher – and so did Bill. His experience started off well. At infant school, Monteagle Infant School in Dagenham, the kids would troop into assembly hand in hand, as children do, to the music teacher's flamboyant rendering of 'Country Gardens', a folk tune arranged for piano by Percy Grainger. "It was my first hearing of 'Country Gardens'. I thought, 'What a funny tune.'"[33]

32 Bill Leader, interview with the author, October 12, 2012.

33 Bill Leader, interview with the author, February 14, 2020.

His school in Mottingham, Elmstead Wood Junior School, was a modern, one-storey building with large windows sited in a Kentish wood and administered by Kent County Council. It was "full of bright young teachers who were burning with the ambition of educating the next generation, and had the ability to do so," in Bill's words. "I had a liberal education, but only briefly, because after about a year, it seemed to me quite arbitrarily, we were told if you lived this side of a datum line, you were sent to another school altogether. London County Council suddenly decided that they should be in charge of my education, and the London County Council school was much farther away."

Calcott Walk, nominally administered by Kent County Council, was close enough to the London County Council boundary for the LCC to take an active interest in Bill's education. Marvels Lane School was old, several storeys high, and very frightening. The outside wall said 'Board School', but it had *INSTITUTION* written all over it.

"A prison, I would say, built in the 1890s in the School Board days, as old as that, and staffed with old-fashioned teachers (most of them, not absolutely all of them), who thought a clip around the ear was as good as a well-delivered lesson. And I hated it because it was a long way to walk, and when you got there it wasn't particularly pleasant, and after that you had to walk the long way back again."[34]

Bill was at Marvels Lane when war was declared in 1939. By then Bill Snr had moved from Vickers to Smart & Brown, a firm of lathe makers with a factory at the Hither Green end of the London borough of Lewisham. Mottingham didn't get bombed, but its elevated position afforded a spectacular panorama of London extending to the Thames basin, where fires were started.

"We had a view through the trees of a particular bit of London. I didn't know enough of my geography of London to know what I was looking at, except when the bombs fell. The burning buildings would throw things into silhouette in a dramatic way and light up the night sky better than they would light up the day sky. And the bit we could see was Canning Town. I suppose that's what we could see when the bombs weren't falling too. It was enough of a display to attract your attention and keep you very worried about what was going on in the world."

34 Bill Leader, interview with the author, October 10, 2014, and the following description of the Blitz.

Canning Town was where Gran and Grandad Leader lived, at 58 Lawrence Street. Aunt Agnes lived around the corner, and other members of Bill's mother's family were in the near vicinity. The Wicks lived nearby. The Leaders and Wicks were shortly to be linked by the marriage of Patrick and Maggie on March 8, 1941, a union blessed by one daughter, Josephine Margaret Leader – Jo – born August 23, 1942.

"For that period, when the Blitz started, the raids came every night as regular as clockwork," remembers Bill. "We lived on the south side of London, which is where, in the main, the bombers came in from. We heard them go overhead. They had a particular strange, throbbing drone. I think they were Dorniers. It started in the evening and went on until the early hours. I'm not sure when exactly, because I went to sleep at some point.

"And even before the Blitz, I remember walking in Kent... Mum and dad used to like to go out on a bit of a ramble on Sundays and take a train to fairly rural parts of Kent, and then ramble to the nearest town and find a pub to have some lunch. We were out there once, a few weeks before the Blitz started, and suddenly there was a dogfight. That was my first experience of the RAF defending us against intruders."

Jo's mum, Maggie, not yet married, went along with them.

"My mum said they were in the pub one Sunday lunchtime and there was this dogfight going on overhead, and as the planes were coming down, she said you could see the pilots' faces, both of them, English and German. She said it was really frightening but at the same time very interesting."[35]

"I went to an air-raid shelter," resumes Bill, "and everybody said, don't lean against the wall. You'll get concussion. I presume because of the shaking of the ground when the bombs hit. So everybody was sitting, leaning forward. It's a lesson I've always remembered: don't lean back against the wall in an air-raid shelter.

"Everybody had a gas mask. You were supposed to carry it everywhere. They came in a cardboard box with a piece of string, so you could carry them over your shoulder. But a person with an eye to enterprise made gas mask cases, with colours, and more elaborate straps, not just string. I suspect the same entrepreneur –with a back-room full of ladies sitting at sewing-machines – made the wallets and covers for ration-books. Every cloud has a silver lining.

35 Jo Pye, interview with the author, December 12, 2013.

"You had to know how to put your gas mask on. You put your chin in first, and then you brought the straps around the back of your head. If you had a baby, mum was supposed to sit there with her own gas mask on, and a container for the baby, and feed air to the baby through its gas mask with a foot-pump. If she had a slightly older child, they would have a Mickey Mouse gas mask. Walt Disney didn't let these opportunities go by. If you're going to transform a gas mask into anything, Mickey Mouse is a good idea."[36]

The safety of Thomas and Martha Louisa was a matter of serious concern, as the Canning Town docks were targeted night after night. A solution was found – they moved in with their daughter and family at 1 Calcott Walk. Other relatives soon followed, anxious to escape the nightly inferno.

"We had a living-room, a kitchen, two bedrooms and a bathroom. It was big enough for the three of us. But then we tried to get in a grandfather, a grandmother, a couple of aunts, a few cousins *and* us. It got a bit squashed at times."

Lil came with her two children, Lily and George. I suspect Kit made up the quota of aunts, but Bill doesn't remember Kit. It was a full household by any standard. They coped with the stoicism that was *de rigueur* for wartime.

> It wasn't so bad in the daytime, but regularly each evening the raiding planes arrived. For distraction we played Newmarket, Chase the Ace or some other simple, any-number-can-play card game, listening to Jack (Blue Pencil) Warner and Joan (Little Gel) Winters in Garrison Theatre, or whatever else was on the radio, and to the creepy, throbbing drone of the twin engined Dornier bombers as they prepared to release their incendiary load on the vacated home of the family gathered around me.[37]

Luckily, Chase the Ace is a game that improves with more players.

"What was I? Nine? So I wasn't terribly conscious of what inter-family tensions there might be. I'm not saying there were any, but if there had been, they would have gone over my head a bit, as long as they didn't end up in outright rows. My mother and her mother got on easily. Grandfather walked down the hill into Mottingham village

36 Bill Leader, interview with the author, October 10, 2014.

37 *If I Remember Writely*, ibid.

and spent most of his days in the Porcupine pub, and then he made his way home."[38]

People have different ways of dealing with stress, and Grandad Leader's way was as good as any: he tottered to the pub and back again. Mottingham had no debris to navigate, unlike the open bomb site that was Canning Town. Grandad Leader was happy, or as happy as was compatible with a general air of grumpiness. Such types proliferated in wartime. 'Old Brown', the subject of a recitation by gypsy singer Albert Smith, was another such:

> *Old Brown sat in The Rose and Crown a-talking about the war.*
> *He dipped his finger into the beer and then began to draw.*
> *"For there are the British lines and there's the German foe."*
> *The potman shouted out, "It's time!"*
> *Old Brown said, "Half a mo!*
> *Do you want us to lose the war?*
> *You've shouted five minutes, I think it a sin*
> *For another half pint we'd have been in Berlin.*
> *Do you want us to lose the war?"*[39]

Eventually, the entire party – grandfather, grandmother, Aunt Lil and the kids – moved to Horning in Norfolk, near Uncle George's village, Irstead. George was in the Navy; it was George's sister who found the accommodation.

> Horning was then a virtually deserted village on the Norfolk Broads; very rural, but a little bit too close to the east coast and the German army on the European mainland. However a popular song of the day reassured us that we were going to hang out our washing on the Germans' Siegfried Line, should the said defence structure still be standing when our victorious soldiers got to it. The music hall duo, Flanagan and Allen, were frequently heard singing this ditty on the radio, but I don't remember the BBC ever playing the reverse side of the record, which was called something like 'If a Grey Haired Lady Says "How's Your Father"'...

38 Bill Leader, interview with the author, March 29, 2018.

39 Albert Smith, 'Old Brown Sat in The Rose And Crown', *The Voice of the People* Vol. 14: *Troubles They Are But Few*, TSCD664.

If a grey haired lady says "How's yer Father?"
That's Mademoiselle from Armentieres

(spoken)
You're going across the sea, my lad
I wish you luck, my son (Thanks, Dad)
You've got a great big job to do
A job that must be done (Yes, Dad)
If you should meet a lady there that I knew years ago (Yes?)
Just give her my kind regards (How will I know?)

If a grey haired lady says "How's yer Father?"
Oh, that will be Mademoiselle
If she smiles and says, "How's yer Father?"
That will be Mademoiselle

If she says, "Parlez-vous, toot suite, tell me do,
How is he after all these years"
If a grey haired lady says "Don't tell your Mother?"
That's Mademoiselle from Armentieres... [40]

A cheeky little number, which, with a nod and wink and a reference back to a favourite song of World War 1, confirmed that it was but one generation since the last time our brave lads had done some dying on the continent.[41]

"And so the house in Lawrence Street," says Jo, picking up the narrative, "was more or less empty, except for when my dad was there, and when Aunt Kit was there. Aunt Kit didn't go originally. Aunt Kit went to Aunt Lou's for a while." (There! I knew I'd heard about Aunt Kit in Mottingham from somewhere.) "They'd had this really bad bombing, and all the people were collected in Agate Street school – South Hallsville was its proper name – and my mother and Granny Wicks were going to the market, shops, whatever, and the air raid

40 'If a Grey Haired Lady Says "How's Your Father"' by Ted Waite, a hit for Flanagan and Allen in 1939.

41 *If I Remember Writely*, ibid.

warden said to them, 'Look, we're expecting another raid.' Well over a hundred bombs had dropped on Canning Town at that point. It was getting on for 300 during the whole period of the war. 'You need to go to the school, shelter there, and we've got transport that is coming to take you away.' And my mother always swore she heard a voice and it said to her, 'Don't go. It's not safe. Go to Mrs Leader's. Mrs Leader has a shelter.' She had a key, I think, to keep an eye on it. 'Go to the Leaders!' And that's where mum, granny and mum's youngest sister – Grandad wouldn't go – went. And the school actually had a direct hit. My mother always said it was a premonition."[42]

The official figure for casualties in the South Hallsville School bombing raid – it took place on September 10, 1940 – was 77, but locals put the death toll at much, much higher. The number was scaled down for the sake of public morale. An error sent the buses that were going to evacuate people to Camden Town instead of Canning Town. When people (who weren't there) invoke the Spirit of the Blitz, the cock-up and carnage need to be remembered too. An estimated 600 people died in the one strike, making South Hallsville the worst civilian tragedy of the war.

"And then a couple of days later Granny Wicks' house got a direct hit, so they had nowhere to go, so they stayed in Granny Leaders'."

This was how 58 Lawrence Street acquired new inhabitants, although the Wicks regarded themselves as custodians, or house minders for the absent Leaders, rather than residents. This explains their protectiveness about the piano in the front room. Only Jo and Lily, as true Leaders, were allowed on it. Jo's generation were taught to play instruments as a matter of course, and Jo's cousin Michael (the son of Jim and Elsie), was the acknowledged musician of the family.

"I suppose the piano went when the house went, because they didn't take a lot of stuff with them. And it was probably really knackered by then, because it had sat in there and been blitzed and god knows what else."

One night, sleeping by himself in the iron air-raid shelter at the bottom of the garden in Lawrence Street, Granddad Wicks woke up, absently noticed how hot it was, and turned over and fell back to sleep. When he woke in the morning, he noticed that the next-door house had been bombed to smithereens. He knew then what he had to do.

42 Jo Pye, interview with the author, March 20, 2018.

"A direct hit would take out not just his house but perhaps the next three or four," explains Jo.

Granddad Wicks gathered every object that was remotely flammable and consigned them to a controlled bonfire. Shellac discs and wax cylinders were fed to the flames.

"It was a sensible thing to do, but a great pity," laments Jo.

"I've never heard anything like that," is Bill's response. "I never saw it or heard it spoken of, but I don't think it involved any of our possessions because my mother would have had all the books and records she could reasonably call hers in Mottingham, or Becontree when we moved there. I don't think she had any possessions in Lawrence Street. My mother would have considered book burning as one of the great horrors of civilisation, but I don't think the others possessed much in the way of books."[43]

Granny Leader was widowed; Thomas Leader died on August 25, 1942.

"His death certificate just says Horning, so I'm assuming that it was in The Round House, the cottage they lived in. He had an operation for cancer a few weeks before, so it could be that they sent him home to die."[44]

In 1940, with no end to the Blitz in sight, the government decided to move Bill Snr's place of work northwards, in line with a policy to relocate works deemed essential to the war effort. Smart & Brown (Engineers) Ltd were to be transplanted from Lewisham to Keighley in

---

43 Bill Leader, interview with the author, March 29, 2018. Upon hearing a rumour that a cylinder gramophone was on the bonfire, he has this to say: "There was one gramophone there which in fact my mother ended up having, which I was brought up with, a big HMV wind-up cabinet gramophone. And Aunt Kit, when she started earning money, bought a radiogram, which was the absolute frontier of technology at that time. You could pile records up on top on a spindle, and they dropped down, hopefully one by one, in order to be able to be played in sequence, assuming you'd put them in sequence. I don't know what happened to that, but that was a big bit of furniture, a radiogram. That was an HMV too, I think. I remember that particularly, because when she bought it she was given some records for free by the record shop and then she passed them on to me, because there were things like Yehudi Menuhin, who was then nobbut a lad, playing some of his tricky little pieces. I forget what else but the Yehudi Menuhins I remember. They were little features, you know, encore pieces. A Capriccio by somebody or other. Lighthearted virtuoso pieces."

44 Jo Pye, by email to the author, March 29, 2018.

the West Riding of Yorkshire, to share the Lawkholme Lane premises of the Rustless Iron Company, or TRICO, specialists in vitreous enamelling of all sorts. It seemed to me, growing up, that every second household had a hearth accessory in the shape of a knight holding a dustpan and brush. Domestic chivalry was born in Keighley.[45]

"So we were all stuffed into a coach one day," says Bill, "rather early in the morning, and spent the entire day on the A1, the Great North Road, which was a bit of a cart-track, making our way slowly to, whatever it's called, this place up there called Keeley, or Keighley."[46]

> Singing was the thing to do on a coach journey. Not first thing, when the day was new and chilly, but later, after a stop for a spot of refreshment, someone would strike up with an old music hall number. My old man said follow the van, and don't dilly dally on the way somehow seemed appropriate. Strangely, there were some songs that always got coupled in a strict sequence: 'Daisy, Daisy' always, always segued into 'She Was a Sweet Little Dicky Bird' for instance. I never found out why. By the time I made my migration north, I was already a veteran of day coach trips. Mum was an active member of the Co-operative Women's Guild, a social/cultural/political adjunct of the Co-operative movement, a sort of Women's Institute for the industrial working class. Both organisations had a touch of jam and 'Jerusalem', a little less jam for the Co-operators, but they were fervent in their singing of the Blake/Parry anthem, seeing in it an affirmation that the world could, and in fact, would become a better place.
>
> At least once a year, usually more frequently, guild members, a bunch of young to middle aged working class housewives who, while their hubbies were working, would

45 Chris Ackroyd briefly worked at TRICO and has mixed feelings about the place. "Once, bored on a nightshift, I climbed the chimney. No-one knew I was there. On the way down the ladder, my legs began to tremble so violently I was in danger of falling off." Chris Ackroyd, letter to the author, October 22, 2017.

46 "Keighley – worth repeating: nobody knows where it is or how to spell it or pronounce it. And yet for decades, especially in a hung-election, Keighley went with the delicate voting balance of the rest of the country – so it became a kind of barometer for journalists on national newspapers once every four years." Chris Ackroyd, ibid.

> decide to 'have a Basinful of the Briny'[47], as a song of the time put it. Southend, Clacton, Dymchurch, Margate and Hastings were popular destinations at the time for people living on the eastern or southern side of London. We didn't call our vehicle a coach; we called it a sharra, our version of char-à-banc, the name originally given to the horse drawn wagons that were used when the concept of day outings had first taken off. Recently I noticed that Collins, the publishers, announced that they had removed 'char-à-banc' from their dictionary, as the word was out of use. I read this the day after I had watched a BBC documentary on The Golden Age of Road Travel, which had frequently resorted to the use of this supposedly redundant word. Who put the Dick in dictionary?
>
> Songs, refreshment stops, subsequent relief stops, a destination stop was the substance and sequence of the outward journey, all to be repeated on the return journey, plus a collection for the driver. I remember two stops the day we were moved north. There was one at Biggleswade, forty miles north of London, where in fact, several of the families on the coach were dropped. The other stop I remember was Doncaster, a further 150 miles on, and my Yorkshire first footing. On this occasion, because there was to be no return leg, the driver's collection was taken.
>
> The North Road called Great was more like a wide footpath in those pre-motorway times. The journey took all day, and a day was forever for an 'are-we-there-yet' kid of ten. Fortunately the Smart & Brown advance staff and the good folk of Keighley had worked well, and on arrival we disembarked at the bus station, where the workforce and their families were quickly taken to the houses of those families who had offered them temporary accommodation.[48]

The Leaders were billeted with a family called Brown, "very nice people", who lived on Thwaites Brow, at the top of a steep, steep hill, whose road was made treacherous by hairpin bends. This part of the

47 'Let's Have a Basinful of the Briny' by Roy Leslie (Eclipse No. 727, 1934).

48 *If I Remember Writely*, ibid.

journey was covered in the Browns' car. Bill was fast asleep, so he missed it. Between them, Bill Snr and Mr Brown carried the limp lump of the boy up the narrow stairs to a makeshift bed in a room at the top of the house.

"The Aire Valley and the Worth Valley join together and in that little bulge is the town of Keighley. It has a reasonably long history of industrialisation. It's surrounded by hills. There's the hill up to Haworth on this side, and Ilkley Moor on that side, and Thwaites Brow is on one of these hills. And we got put into this house late at night, and I was put to bed. And then in the morning I could hear strange voices, speaking in a strange accent. I got up and went to the window, and drew the curtain, and there we were on the edge of this sheer drop, with the Aire Valley going in both directions away, and Ilkley Moor, or Rombalds Moor, in front of me. It was the most spectacular panorama I'd ever seen, and one of the most impressive experiences of my life. It's left me with a great desire never to be too far away from the Pennines, and preferably closer than Middleton."[49]

> My first impression was that a small, provincial, industrial town had been placed amongst impressive hills, taken back a little in time, and peopled by actors in a costume of flat caps and clogs, who had all adopted the ee-by-gum accent used by comics. I crossed the north-south divide.[50]

Bill Snr became a shop steward at Smart & Brown Ltd, a position he held during the rest of the war. When the arrangement with TRICO came to an end – not immediately upon the end of the war, but around the late forties – Smart & Brown moved to its twin bases in Durham and Biggleswade and continued to make world-class precision tools until its demise in the early eighties. Bill Snr remained in Shipley and found employ at the Metal Box Company (currently Novar plc), where he remained until retirement.

Because war brought a shift in traditional gender roles and women were required to do work that was formerly the preserve of men, Lou became an engineer. She worked for a Shipley company unblushingly called Parkinson's Perfect Vices, and later at the GEC (the English subsidiary of General Electric, the US company), which operated from a large ex-printing factory on a hill outside Bradford.

49 Bill Leader, interview with the author, March 2, 2012.

50 *If I Remember Writely*, ibid.

"Here's a picture of those smart and bright lads at TRICO, 1945. There he is, that's my dad..." Second from end right, second row (Bill Leader collection)

"She worked there for quite a long time. They were the two major ones that I remember."[51]

Alex Eaton, Bill's friend and mentor, mentions a third in *Dot and Carry One*, a memoir written for his grandchildren.

> You may remember Louise [*sic*] Leader, a powerful woman who during the war worked at Avro aircraft at Yeadon, cooked and washed and ironed for her son and ailing husband and still a Shop Steward ["I'm not sure she was a shop steward, but I would rely on Alex more than me anytime"], a seller of the *Daily Worker* and initiator and organiser of the yearly Christmas 'Bazaar' to raise funds for it, for it had no revenue from advertising of course. She joined others at Avro at 6am to sell the Worker to the departing nightshift before going in to work themselves.[52]

Two reactions from Bill:

"I'm not happy about Mum selling the *Worker* at Avro's. I don't remember her working at Avro's (but her younger sister Kit did).

---

51 Bill Leader, interview with the author, June 27, 2018.

52 Alex Eaton, *Dot and Carry One*, eBook; pp. 1,426-1,428.

The Avro referred to was not at Woodford (which was in Stockport, many, many miles and a mountain away from Shipley) – it was at Yeadon. The building is now part of (or near to) Leeds/Bradford Airport. I think that Alex got the *Worker* selling right, but the place wrong."[53]

and...

"The detail about selling the *Daily Worker* to the departing night shift before the start of her own working day is true. I mean she wouldn't have done that every morning, but there would have been special occasions when the need arose."[54]

The *Daily Worker* was the only paper Lou trusted. In January 1941 it was banned by Home Secretary Herbert Morrison for its provocative stance on the war. Its readers rallied around, and Lou, as ever, was generous in her support.

Immediately the printing and distribution went underground. "Mum was involved in the illicit distribution, made easier in Bradford by the fact that Ukrainian-born, Canadian-naturalised Bert Ramelson was managing the Bradford Marks and Spencer's branch, and the shop became the local distribution point for the paper."[55]

---

53 Bill Leader, annotation, Jul 4, 2020.

54 Bill Leader, interview with the author, June 27, 2018.

55 Bill Leader, annotation, September 3, 2018. The *Guardian* editorial, Jan 22, 1941, sets out the establishment's case: "No one likes the idea of the suppression of a newspaper even during a war, and least of all the suppression of a newspaper that is the sole organ of a legal political party. It is one of the last steps an Executive should take. Yet no one who has read the *Daily Worker*... can doubt the extreme provocation they have given... The *Daily Worker* began the war as a supporter of resistance to Hitler; it changed its tune when it found that Stalin wanted to be friends with Hitler. Day after day since it has vilified the British Government and its leaders to the exclusion of any condemnation of Hitler. Nothing that has happened in this country has been decent and right. Even when the United States increases its aid this is denounced not as something to be welcomed but as a malevolent exercise of wicked Yankee capitalism. More recently the paper has largely devoted its columns to derogatory accounts of Service conditions on the one hand and to the encouragement of agitation among munition workers on the other. This might be excusable if the motive were honest, if it were really desired to help the country in its struggle to keep democracy alive in Europe. But the *Daily Worker* did not believe either in the war or in democracy; its aim was to confuse and weaken. We can well spare it."

> In her 'free' time at home she knitted, sewed and made rugs for sale at the bazaar. She made the black tab rug[56] you've seen in our caravan. No wonder her face was grey with fatigue.

"I could be completely insensitive about how fatigued she might have been, or she might have deliberately not showed it. I remember her as a cheerful smiling person. I don't remember her 'grey with fatigue'. But he's not a fantasist, our Alex."

> Her tongue was crisp in defence of the Party and its ideals as she fought for better working conditions and equality for women in the factory where of course for the same work women were currently paid less than a man… Louise was outstanding, but she was one of many.

"Very good. Good old Alex."

Worse things happen in wartime than a full house, an epiphany on top of a hill in the West Riding, an interrupted education (Bill had not attended school since the onset of the Blitz) and having a mother who was one of many. Karl Dallas, Bill's near contemporary, was evacuated to Northumberland and wrote to his mother saying he would rather die in an air raid in London with her than be safe in the countryside without her.[57] It was an early example of his effectiveness with a pen. Naturally, his mother gave in and Karl was restored to Acton.

Did Bill and Karl experience the pure joy known to John Foreman? (Bill, yes; I'm not so sure about Karl.)

"Kids had a lovely time," says John. "The ones who were evacuated weren't so happy, but those of us who were here, yes. School was a joke! I used to collect shrapnel. It was a great thing, first thing in the morning, going out, seeing what had been bombed, digging out the bits and pieces, finding bits of guns. We were able to play about on bomb sites. Oh, it was very jolly."[58]

The war was more sombre for grown-ups. Silence became a part of self-preservation. Marie Little and I are discussing the possibility

56 "Tab rugs were Mum's speciality. She made many of the size that would go in front of a roaring fire. She once made one that was a very long stair carpet. It took ages for her to make." Bill Leader, annotation, June 27. 2020.

57 Stephen Dallas, address at the Karl Dallas Memorial, June 30, 2016.

58 John Foreman, interview with the author, October 7, 2017.

that Ewan MacColl and Paul Graney were deserters. It's circumstantial evidence admittedly, but why should the Second World War be so conspicuously absent from the memoirs of both?[59]

"See, lots of people didn't talk about the war. Dad never talked about the war. They didn't mention the war because it was just too painful. It was almost like a social taboo. I remember my dad, he was on the convoys, and my mamma said, you never talk about it. The only time he said something to my mother was like, one of the ships had been blown up, and there was lads in the water and they were shouting for help, and he said we could have stopped and picked 'em up, and they weren't allowed to stop and pick 'em up. And that always stayed with my dad. And that's the only one thing that he ever mentioned to me mam about the war. It was like it was too painful. So many people wouldn't talk about the war because the whole experience was an awful bloody trauma."[60]

Too painful to talk about and too traumatic to forget; that's deep-rooted psychological damage in a nutshell. Artist-designer Jim Boswell was working for Shell when he received the call-up.

"He became a radiographer, because that's what they decided he should do. And he got himself about 18 months in Iraq, and while he was in Iraq he did a huge amount of drawing."[61]

Boswell's drawings and watercolours of petrol dumps, lorries, and the general camp life of "squaddies and wadis"[62] convey the monotony of life in the desert. Artist Russell Quaye was less fortunate and saw what is euphemistically described as 'action'. His ship was torpedoed on its way to Malta.

"He was one of a dozen survivors clinging to a coal-chute that had become detached, floating around in the Med, in total shock, with weaker ones dropping off and disappearing every so often," says Hylda Sims, his future partner in life and music. "He was on the last hospital ship to leave Tobruk before the Germans took it. I think he was in quite a bit of a mess when the war ended."[63]

59 Respectively, Paul Graney, *One Bloke* (Bluecoat Press, 2011) and Ewan MacColl, *Journeyman* (Sidgwick & Jackson, 1990).

60 Marie Little, interview with the author, June 24, 2013.

61 Sally Shuel, interview with the author, October 17, 2017.

62 'Boswell Remembered,: A Memoir by James Friell', op. cit.

63 Pete Frame, *The Restless Generation: How rock music changed the face of 1950s Britain* (Rogan House, 2007), p. 114.

Hylda, a child, was closer to the John Foreman model. War meant an end to a peripatetic existence in a caravan with her parents, a runaway couple, and fulfilment at Summerhill, the progressive boarding school, evacuated to Wales for the duration.

Here's Hamish Henderson indulging in some Hollyrood-style heroics with the Highland Regiment on the beachhead at Sicily in July 1943:

> He requisitioned a large white stallion from a nearby farm and rode down from the dunes; as one veteran remembered half a century later, he 'wis gettin' it to rise up on its hind legs, like in an oil painting! And the boys were all cheering'. "Splashing through waves 'wi' his bonnet akimbo," Henderson had Bonnie Prince Charlie in mind.[64]

Yet the writer of 'D-Day Dodgers' knew something about the pity of war. Incidentally, Hamish Henderson personally accepted the surrender of Italy from Marshal Graziani on April 29, 1945.

Ewan MacColl's major contribution to the war effort was the scabrous, disgruntled song 'Browned Off', a real toxic for the troops.

Of Bill's other associates, Guyanese-born Cy Grant was allowed to join the RAF after the high mortality rate among crew members prompted a revision of the colour bar. Cy was still banned from being a pilot, however, and was serving as a navigator when he was shot down on his third bombing raid over Germany. Two of the crew perished; he and the pilot spent two years in a POW camp. At war's end, the prisoners were force-marched to Luckenwalde, a holding camp 52 kilometres to the south of Berlin.

"Then one morning we woke up and there were no guards. The Germans just disappeared. They fled. They knew the Russians were there."

It was not the end of his troubles.

"I was once sent out on a foraging party to bring back food. I thought we were going to raid some German shop and steal food, but [*chuckles*] we had to round up cattle. Now how would I know how to catch a cow?"

It was winter and the hard, snow-covered ground was littered with corpses.

64 Patrick Wright, 'His Bonnet Akimbo', *London Review of Books*, Vol. 33, No. 21, November, 2011, p. 31. Wright is reviewing *Hamish Henderson: A Biography* (Polygon) by Timothy Neat.

"My pilot was a Canadian guy. He didn't know anything about it. He got a big piece of wood, approached a cow, hit it on its head; the cow fell down. He took out his penknife and cut its throat. And that was even worse, because a cow is a heavy thing, and to drag a dead cow and pull it into a truck was impossible. But eventually, somebody had the idea, why don't we get a halter? So we went to the farmer, who reluctantly gave us a halter, captured the cows and made a ramp and caught about six or seven… But we had to go into the same part of the truck as these cows, and the cows were petrified now. And I can remember sitting right at the end of the truck, and the tail of a cow was waving over my head, and suddenly I got a blessing… which was the most disgusting state of affairs."[65]

Street fighter and bookseller Jack Firestein, as a member of the Royal Fusiliers, was downed by a bullet at Anzio. He was taken to an Italian hospital, and thence transferred by train to Germany. He was urged to throw away his identification tag, the material evidence of his Jewish ethnicity. He complied, but clearly resented it, because when an SS officer came around shortly afterwards taking names, he made no attempt at concealment.

"Jack Firestein," he said.

This gave the SS man pause. He pondered.

"Firestein is not an English name."

"Yes, it's English," Jack returned. "There's Brown, Smith, and Firestein."

The SS officer shrugged and moved on.[66]

Max Alexander relays the wartime experience of his parents, Joyce and John Alexander (these people have some small part in Bill's life, and we'll be meeting them all). In the last months of the war, Joyce was an army nurse on a hospital train from Hull to Leeds. She felt largely helpless, as all she could offer the casualties was a 'vim tin' of disinfectant powder to put on their wounds, and soft words to soothe the trauma.

But Joyce was a member of the CPGB and agitating for women's rights within the force. She was dishonourably discharged by way of a detention camp where inmates were made to load and unload US military lorries.

65 Cy Grant interview Windrush Foundation, www.youtube.com/watch?v=SDT-HJCpSKQ.

66 John Foreman, interview with the author, July 9, 2016. Firestein was the subject of Chris Reeves' documentary, *Only a Bookseller*, in 2009.

"She was pregnant with my sister and Dad did the right thing and married her, signing up as a professional soldier in order to support his new family. He was in the middle of an exam in the Grand Hall at Eltham Palace (the HQ of the Education Corps) to gain a librarian's qualification, when an officer came and asked him to report to his commander, who told him that he was being immediately discharged and that he would know the reason why."

"So Dad was chucked out of the army for marrying Mum."[67]

And what of Ray Kinsey, the ex-boy soprano? Kinsey, with the fearlessness he had previously reserved for tackling Handel arias, distinguished himself as a squadron leader and bomber pilot. After a series of traumatic stress episodes (or nervous breakdowns, as they were commonly known), he was relieved of combat duties and transferred to Canada to become a pilot instructor. Oh, for a closer walk with God, indeed.

67 Max Alexander, email to the author, April 24, 2020.

Come on boy and take your place
Among the men who serve the trade
IAN CAMPBELL, 'APPRENTICE SONG'

IV

Thistlethwaite, the Road Not Taken

Another historical eminence besides Karl Marx now entered Bill's life. Sir Titus Salt was a Victorian entrepreneur who made a fortune from alpaca. The story goes that Salt was in Liverpool buying imported wool and noticed two forlorn bales of alpaca lying neglected on the dockside. Salt bought them cheap and carried them to Bradford. Alpaca – from the llama-like South American goat of the same name – has longer and silkier fibre than wool and is plucked rather than sheared. It is much harder to manipulate. Sir Titus relished a challenge, however, and ordered his engineers to build looms capable of handling the unusual fabric. Alpaca permitted a new softness in clothes; soon fashionable women were wearing alpaca dresses and fashionable men were sporting alpaca waistcoats.

"Salt was the man who solved the problems in manufacturing alpaca. He cornered the market, like clever people do. It became a very popular fashion item."[1]

Rich beyond avarice and fired by paternal benevolence, Sir Titus set out to build a model village for his workers. He picked a greenfield site in the Bradford area, adjacent to Salts Mill, the largest of his mills.

1 Bill Leader, interview with the author, March 15, 2012.

It was next to the river and had good communications with a nearby railway and canal.

With the chutzpah special to Victorian industrial magnates, he called the new town Saltaire, combining his name with the name of the river.

As a loyal subject, he consecrated the two principal thoroughfares to his queen and her consort. With Victoria Road and Albert Road out of the way, Sir Titus could get personal. Caroline Street was for his wife, and Fanny Street, William Henry Street ("a good street that," says Bill),[2] Jane Street, George Street and Mary Street (which continued as Upper Mary Street) were named after Sir Titus' offspring. Titus Street is not ego-driven but named after another son.

And whereas nearly everybody was a Methodist, Sir Titus was a Congregationalist, and decreed a Congregationalist church on the banks of the Aire. Saltaire United Reformed Church is atypical for the region, being round and elaborate as opposed to square and plain and is clearly modelled on a Roman temple just to rile the anti-papists. Yet he was a good employer and made sure Salts Hospital was close to Salts Mill on Victoria Road, so injuries sustained by his workforce could be promptly treated. Victoria Hall, also on Victoria Road, housed a library, gymnasium, rifle drill-room, fencing room, armoury, chess room, laboratory and lecture theatre – all for the greater good of the community. Distinguished speakers such as John Ruskin, Benjamin Disraeli and David Livingstone packed the lecture theatre. Sadly, Charles Dickens died before he could fulfil his booking. Later, Victoria Hall would host the annual *Daily Worker* bazaar, where Lou would sell her handmade woollen rugs. Still later it hosted fund-raising events for the 1984-85 Miners' Strike.

It was not only a model village, it was also an ideal village, organised along lines approved by Sir Titus Salt. The sale of alcohol was strictly forbidden.

"I can see why you might be against strong drink if you're a mill owner and you've got a lot of working people using heavy machinery around you. You don't want people falling into the looms, and you particularly don't want people falling into the carding machines. That really is messy."[3]

2 Bill Leader, interview with the author, March 2, 2012.

3 Bill Leader, interview with the author, March 15, 2012.

Saltaire is virtually dry a century and a half later, although there is a bar on Victoria Road tellingly called Don't Tell Titus. Because drinkers must go somewhere, the road just past the town boundary was far-famed as a pub crawl.

"The pubs were cheek by jowl and interestingly, because this was before all the brewers started gobbling each other up, there was a huge variety of breweries represented in the pubs. And so if you started in Low Well, Shipley, and worked your way up to where the ban started at Salts Hospital, you could have a delightful experience of inebriation, and very tasty too. Most of the brews have disappeared long ago. I don't think Shipley had a brewery. Bradford had a couple. Halifax had seventeen, I read somewhere. Halifax water was good stuff."

> Much of the housing around this part of the Pennines is pretty impressive. Not the slum stock of course, but the use of the local millstone grit and an attractive vernacular building style has meant that the many houses built between the wool boom and the nineteen twenties were – and indeed still are – handsome and pleasing. And as me and my mum went round, we found that many of them were not just handsome but also empty. This amount of quite large, empty, housing had proved lucky for the authorities when earlier on, in the spring of 1940, places had quickly to be found, somewhere, anywhere, for those troops who, a few months earlier, had managed to get away from the disastrous defeat of Dunkirk. And it wasn't just the fleeing British Expeditionary Force that had to be billeted back in Britain; daily there were loads of newly conscripted men to house. Many of the empty houses we viewed were empty because troops had just been moved out to be posted elsewhere, and your average squaddie does not tidy up before he leaves. We tried not to be too depressed by what we saw. We knew we'd find something liveable, and eventually did. Very nice, if a bit large for us, an oldish terrace house, just outside the model village that the textile entrepreneur Sir Titus Salt built for the workers, who worked at the mill which he had built on the banks of the river Aire and had called, with all the modesty he could muster, Saltaire.

> We were now settled into Park Grove, a stubby, unmade-up cul-de-sac which contained, on its west side, a single terrace of 16 Victorian houses over-looking the vicarage garden of the church of St Peter. At the corner, on the main road, was the Gaumont Picture House, which showed films a little later than the Bradford cinemas, and slightly more damaged. I remember the first time I went there, I was charged 3d (three old pence, 0.0125 of one pound sterling) to get in. In return for my 3d I got a ticket with 2d printed on it, because the price had only recently gone up and they were still using the old tickets. The proximity of the cinema and the reasonableness of the admission charge, kick-started my love of film – by the way, the last time I went to the pictures, it cost me about 400 times more than 3d.[4]

Bill could at last resume his interrupted education at Albert Road Junior School, where he repeated a year. It was at secondary school, the "pompously named" Shipley Selective Central School, that he plummeted to the bottom of the class.

> It wasn't helped by the headmaster being a sanctimonious bully, who bore a disturbing similarity to the pen portrait in a book I had, called *When Justice Faltered*.[5] The portrait was of Rev. John Selby Watson, a headmaster, known in his day for thrashing his pupils; beating his wife's brains out, and getting away with it. My headmaster stopped at murdering his wife; I know this because she taught in the local high school, but he was certainly given to knocking his pupils about. He would call the entire school to be prayed over and ranted at, at least twice a day, frequently pointing out to the pupils that, in these times of war, the priority of allegiances they should hold was firstly to their school, secondly their country, and then to their family. Then, after asking for the blessings of gentle Jesus meek

4 *If I Remember Writely*, ibid.

5 Wikipedia is surprisingly charitable toward uxoricide John Selby Watson but Bill has low tolerance levels of bullying, mentally disturbed headmasters. His source, *When Justice Faltered: A Study of Nine Peculiar Murder Trials* by Richard S. Lambert, was published in 1935. Watson, by the way, didn't get away with it. His death sentence was commuted to life imprisonment and he died in Parkhurst prison twelve years after the dastardly deed.

> and mild, he would scream at any lad who in the course of the homily, had slumped a little from an upright position – "*Stand up boy, or I'll knock you down!*" – with such force that he dislodged the top set of his false teeth. He was no good as a headmaster, but I came to discover that he was an excellent maths teacher.[6]

"I'd not had much schooling for a while because of the Blitz.[7] I didn't go to school in Keighley at all. When we eventually found a place in Saltaire to live, I'd been off school for a while. I was too old to re-sit the County Minor. They had a central school, which was a school that selected its pupils. It scooped up some of the people that the education system didn't consider complete write-offs. If you could put a couple of words together, they would let you in. As opposed to the senior school, which was the slum school of the area (of any area, I think). They became secondary moderns after the '44 Education Act. So I went to the school where the half-bright lads and lasses went. That was part of my education. The other part of my education was the youth club.

"Shipley Youth Club was the non-sectarian youth club in the area. Youth clubs were the big things for keeping kids off the street, particularly during the war, when everyone had bright-eyed ideas about the positive aspects of education and how post-war society would be better. The Windhill Industrial Co-operative Society had a youth club and so did various religious groups. Shipley Youth Club was a good one, council-supported, and ran by a lad who was the main electrician at the local motorcycle factory, Scott's. We were very proud of the fact that Scott's Motorbikes were made in Shipley. And Peter Jackson, the electrician, was a very bright lad: a gifted youth leader and a cultured fellow."

Peter Jackson got the kids involved in making what he called 'radio programmes'.

"They were more to do with performing and script-writing than the actual technicalities of broadcasting. We wrote scripts and acted the parts around a primitive microphone. Everyone else was locked up in another room and we fed our efforts via a loudspeaker. That was pretty

---

6 *If I Remember Writely*, ibid.

7 Bill Leader, interview with the author, March 2, 2012, and segues into an interview with the author, March 15, 2012, for the Peter Jackson discussion.

advanced technology in those days. It was possible because Jackson was more than a just a technician. He was a very literate and cultured fellow. He was a liberal-minded lad, with broad-based interests and capabilities. I suspect he was gay, but at the time we didn't know the word (it probably didn't exist at that point). I'm not saying we didn't know about these things. There were nudgey jokes about "funny fellas". But looking back at Peter Jackson, he probably was gay. It had possibly isolated him from some aspects of social living and turned his mind to concentrate on others."

"Sublimation?"

"Probably. He was a good-looking fellow. And he had all the social graces. I don't know what happened to Peter Jackson. He got another job somewhere farther south, I'm not quite sure where, and we got another youth leader who didn't have the same charisma. He was OK, but there was a bit of magic about Peter Jackson.

"I got my first inkling of anti-Semitism there. I had a mate. We used to write scripts together and we worked on various ideas. He came around to our house once. He said, 'There's some people going around the Youth Club saying that Peter Jackson is a Jew.' My mother and father were in the room at the same time and we didn't quite understand what he was saying. 'Oh, I didn't realise.' [In a puzzled tone.] He said, 'But they can't go round saying things like that.' 'Why not?' He says, 'Well everyone knows the Jews are the cause of all the trouble in the world. They own all the banks. They're just nothing but a lot of trouble.' Obviously, Peter Jackson was not a lot of trouble, so therefore he couldn't be a Jew. And nobody should be able to say he was. It was the first time I'd come across anti-semitism in that overt way. I'd heard jokes, the slight references of money-grubbing, but not actually accusing Jewry of all the world's ills. That was a shock. I suppose it shouldn't have been.

"I suppose Peter Jackson had an influence on me. I got a hint from him of how people could organise and run an organisation. Everybody respected him. He seemed to achieve what he set out to achieve, in so far as we knew or were aware of his wishes. He was probably frustrated in many of his efforts. He had an adult-imposed committee around him, which included Mr Richardson, a dispensing chemist with a shop on Bingley Road, who was gaunt and cadaverous. Rather like Tony, his son, he was a thin, thin man.

"There was a manufacturing chemist called Carter, and he was quite the opposite. He was short, and portly, and looked as if he liked a drink. Whereas Mr Richardson was a very aesthetic man. So there were two of them, both chemists. Completely different people, I think. Deeply right-wing, conservative people."

Tony Richardson the director, as you would expect (and hope), is more generous: "Chemists [were] the poor man's doctor... Father's gentleness and patience were extraordinary... He loved people. He related to them. They in turn related to him."[8]

> Back then it was normal for working class kids to leave school at fourteen. During my school days the 1944 Education Act was passed by the wartime government. This raised the school leaving age to fifteen, improved access to grammar schools, changed the name of the sink schools to Secondary Modern and changed the name of the sort of school I was attending to Technical Schools. The technical school thing never really happened. My half-way house type of school took kids to the School Certificate examination. Most of the kids left at fifteen, the boys to go into the textile mills, or the engineering factories, the girls to go into the mill, there to await marriage and motherhood, preferably in that order.[9]

Poor old Rab Butler came in for flak from right and left. "Butler, of course, is sub-human", was the view of Evelyn Waugh. From the opposite end of the political spectrum, the topical songs in *Sing* roast him over the coals. The election supplement (vol. 2, #1, April 1955) mentions him a few times: *Mr Butler's budget made it easy for the boss...* and *Old Mr Butler is worse than a thief...*[10] These allude to Butler's record as Chancellor of the Exchequer. Yet posterity judges him an outstanding Education Minister and a great lost Tory leader. The Education Act of 1944 is named after him.

The Butler Act is arguably the most progressive piece of social legislation enacted by a Conservative government. (What, do you

8 Tony Richardson, *Long Distance Runner* (Faber and Faber, 1993), pp. 10-11.

9 *If I Remember Writely*, ibid.

10 Evelyn Waugh to Ann Fleming (18 July 1963), *The Letters of Evelyn Waugh*, Mark Amory (ed.), (Weidenfeld and Nicolson, 1980), p. 610; cit. https://en.wikiquote.org/wiki/Rab_Butler. The song quotes come from 'The Election Stamp' by Eric Winter and 'On Top of No Smoking' by John Hasted, *Sing* vol. 2, no. 1 (April 1955).

Shipley Youth Club, in a posed photograph for the Shipley Guardian, 1947. Joyce Tetley, second left, top; Bill in profile, third right, bottom. The fellow at far left, playing Ping Pong, is Peter Jackson's successor (collection: Bill Leader)

"That's me and my dad. I don't know how old I was. I was probably quite young" (collection: Bill Leader)

know another? Like the 1980 Housing Act?) Whether they would admit it or no, it was a step towards the goals that activists like Bill Snr and Lou had been fighting for: raised school-leaving age, decent education, practical means for social mobility. The Act tackled inequality by introducing three categories of school – grammar, secondary modern and technical. Children were allocated to each based on an examination at the age of 11, the '11 plus'. Bill, aged 14 in 1944, was just the age and in the right situation to fall through the cracks of the Butler Act.

"I don't think I was put on this earth to be academically bright, but having been shifted around from school to school, I didn't have a natural tendency to do well in these things."[11]

Matters were not helped when a splinter in a snowball infected his hand with septicaemia.

> The winter of '46-'47 was terrible. The war had been over a little over a year when the snow hit. Road and rail blocked. No coal for home or factory. Domestic electricity interrupted. They even had deep now in the Scilly Isles, a place better known for its palm trees. We kids had a ball, several in fact, and they were all made of snow. But three months of it was too much. If only you could have the snow and the snow fights in decent weather. If only the bloody stuff would go away without putting the land under water.
>
> An accident during a snow fight sent my hand septic and I was rushed to hospital, with a raging temperature and the possibility of amputation. I was in Bradford Royal Infirmary for nearly a month; around me a mixture of appallingly wounded soldiers, still attempting to recover from severe battle damage, and loads of weather victims, who had slipped over and broken a limb or two. My raging temperature eventually abated thanks to the use of penicillin, which had only newly become available to civilians. There followed weeks attending the Outpatients department, with my left hand damaged (it still is). The upshot of it all was that I wasn't up to taking my School Certificate on the due date, and I made a mess of it when I eventually took it in the Autumn. The School Certificate was an all or nothing

11 Bill Leader, interview with the author, March 2, 2012.

> exam, and although I passed some subjects, I needed to pass everything to gain the qualification. I left school uncertified and started work at the back end of 1947.[12]

Now that it was time to be earning, the question arose of a fit occupation. What trade would best suit Bill's talents?

"I wanted to get into audio. I had these little fantasies that I was working in the sound part of a film studio. I don't know why. Other kids wanted to drive the Flying Scotsman. I was an enthusiast for recordings since my mother's collection of Gilbert and Sullivan and other things. I learnt my music from records in the main. But why the audio industry as an industry, I don't know. But it was there. And the only people I knew who dealt with audio were the people who ran little shops which sold electrical equipment, and which would let out these strange megaphone things for the tops of vans and which you could stick in the local sports field to announce the results of races, and other events that needed some primitive PA. 'Primitive' being the only version of PA that was available. Our local man was a fellow called Thistlethwaite. Where did he come from? Not from the south, certainly. He had a shop just down the road on the way to Bradford. He was a little more enterprising than the others. He actually offered a direct to disc recording service. (He wasn't alone, but they weren't that many around.) You could go in and speak into a microphone and come away with a little aluminium disc of a song or a message. He also built a tape recorder when tape recorders were very rare indeed."[13]

> Frank Thistlethwaite developed a pocket radio in 1928 which so interested the Bradford Police that he joined them in 1930 to establish their radio communication system!
>
> In 1938 he set up a private recording studio, using a disc recorder, which he operated part-time. During the war, he spent much time in research work and, after the war, reestablished his recording studio. Then in 1950, using a discarded EMI BTR1 tape deck, he designed and built a tape recorder specifically for adding full-track synchronised sound to 16mm films. (These ran at 16 frames per second (equivalent to 7.2ips) whereas 35mm cinema films usually ran at 24fps (18 its) giving, pro rata, a greatly improved

12 *If I Remember Writely*, ibid.

13 Bill Leader, interview with the author, March 15, 2012.

> frequency response). In 1950, he established Excel Sound Services Ltd of 49, Bradford Road, Shipley, Yorkshire and took orders for his new deck, which he launched in the summer of 1951 as the 'Excel'.[14]

"He was into electronics, along with a lot of other people in that area. It was a hotbed of electronics. Just down the road was Wharfedale Wireless Works. That was in Idle, where you'll find the Idle Curiosity Shop and the Idle Workingmen's Club. You'd also find Jowett Motors, which is a famous motor car of unusual breed. And at Brighouse, which is not that far away, there was a fellow called Sugden (another local name, I think), who was producing high quality amplifiers and record players. He designed a high-quality pick-up that was better than the others, and he also developed a machine for cutting discs, long-playing records. In fact, Sugden was among the first to develop and demonstrate stereo disc recording in this country."

> In the nearby village of Idle, for instance, in addition to a place selling antiques that called itself the Idle Curiosity Shop, there was a modern looking building that had a hoarding nearby saying ambiguously, 'Wharfedale Wireless Works'. The building belonged, in fact, to Jowett Cars Ltd: the Wharfedale works were in an adjacent, more modest shed-like structure, doing pioneer work with high quality loudspeakers. Wharfedale's owner, Gilbert Briggs, not only ran this workshop, but also wrote very accessibly about audio. He was also good at booking places like the Festival Hall, selling tickets, and filling the auditorium not just with an audience but also with sound from a hearing aid earpiece which he played through the large horn of an old phonograph. Hi-fi audio was in its infancy in Britain and the pioneer, entrepreneurial, enthusiastic geniuses who were involved seemed to do their best work in their garden sheds. Like Arnold Sugden, working in the not very far away town of Brighouse, producing his innovative Connoisseur line of high grade audio gear, who later developed a ground-breaking, affordable disc cutting machine, turning out stereo long playing disc recordings

14 Barry M. Jones, *A Guide to British Tape Recorders – fifth edition* (2005, Barry M Jones), p. 300.

> some years before the industry big boys even got around to realising which way the future lay.[15]

"So there was a lot of activity in that field of endeavour. Not that I was terribly aware of it at the time. But I did think that working at Thistlethwaite's might help me in my desire to get close to recording."

This begs the question, why such a proliferation of audio expertise and activity in West Yorkshire? Why do the pioneers of this cutting-edge technology have names like Thistlethwaite and Sugden? The former is Middle English, and means a meadow overgrown with thistles; the latter, a popular name in the old West Riding, is Anglo-Saxon and derives from the Olde English 'sucga', 'sparrow', appended by the pre-seventh century 'denn', meaning swine or pasture. We might hazard that the audio industry developed because of the long-established textile and engineering industry in the area.

When Bill mildly mentioned he would like to work at Thistlethwaite's, Lou exploded "*No son of mine is going to work in a shop!*"[16] When she had sufficiently calmed, Lou explained her reasoning:

"You don't want to work in a shop. Conditions are terrible in a shop. They exploit you, and there's nothing you can do about it because you're there all by yourself. But in a factory when they exploit you, which they inevitably do, you can band together and strike for better conditions."

Bill Snr concurred. He urged his son to work in a factory because "he saw the future as organised labour, and organised labour implies large collections of people, and that implies a factory."[17]

> It might sound as though my parents were being doctinaire, but consider the context: it was immediately post-war; the electorate had shown clearly that they had every intention that the unemployment-ridden working conditions of the thirties would never return; militancy was all around. The rank and file had fought, died and won the war: winning the peace was a piece of piss.[18]

"And that's what I did in the end. Went to an inappropriate place and took an inappropriate trade. I was apprenticed as an electrical fitter and spent five fairly fallow years. But it wasn't a complete waste of time,

---

15 *If I Remember Writely*, ibid.

16 Jo Pye, interview with the author, March 20, 2018.

17 Bill Leader, interview with the author, February 13, 2015.

18 Bill Leader, annotation, July 19, 2020.

because I spent five years in a factory, on the factory floor, which is a useful experience and of course is now quite rare."[19]

The firm was the Airedale Electrical and Manufacturing Company.

"They made switch gears and starters for motors. They'd just moved to this site in Apperley Bridge from a place in Cleckheaton, which is Huddersfield way. Most of the people there had actually moved with the factory, and they made the journey across Bradford from Cleckheaton to Apperley Bridge. Which was fairly unusual. At that time people didn't really travel to go to work, outside of London. And most of them had previously worked for a firm they called the Scandi. They all reminisced about the Scandi. I think it was called Scandinavia Mills,[20] and they were deeply into the manufacture of asbestos belting. So all the people I worked with had recently, not purposely but incidentally, moved out of the asbestos industry into the slightly healthier electrical industry. In those days, of course, we knew that asbestos was a good thing. It was fire-proof, and useful for all sorts of things. You could eat it, breathe it in. You could do what you liked with it. It was one of God's little gifts. I wonder how they ended their lives, most of those people? I suppose all of them went to an early death. But that was after I left.

"The apprenticeship was a bit of a farce. You didn't learn anything very much. You were just cheap labour really. You were on the production line but paid a fraction of what the adults were paid. They had no scheme for learning, but what they could have taught I wasn't really interested in or inclined to assimilate. We got day release to Bradford Technical College. One day a week we went for some formal education for a qualification (a National Certificate I think it was called). You could become a member of the Institute of Electrical Engineers if you were a really bright lad. So being an apprentice didn't really help. I had a little bit of money in my pocket to allow me to do other things, I suppose. *Like buy records.*

"And I met lots of interesting people. There were all these old fellows around with advice to young lads about life in general. There were things you believed, and things you were a bit sceptical about. One old fella said, 'You know how to get twins, don't you?' He says, 'Don't touch the wife…' (I didn't have a wife to touch) '…Don't touch the wife for a month, and don't do anything else. Don't have a wank or anything

19 Bill Leader, interview with the author, March 15, 2012.

20 British Belting and Asbestos Limited, manufacturers, of Scandinavia Mills, Cleckheaton, Yorkshire.

like that,' he says. 'And then the next time you fuck the wife,' he says, 'it will be twins.'

"That's a useful bit of information for a young lad to have."[21]

Apperley Bridge takes its name from the old bridge over the River Aire on Apperley Lane, and is just downstream from Esholt.

"I don't know if you know Esholt at all?[22] Esholt has got two reasons for being famous. One is that it was the original location for *Emmerdale Farm*, the Yorkshire Television soap, and the other is it's the home of Bradford Sewage Works.[23] At that time its proud boast was that it was the only sewage works in the world that made a profit. They made a profit because most of the world's wool-combing factories were in Bradford. When they were cleaning their wool, about twice a day, they let loose all this fat-ridden water by washing the fleeces, and it flowed down the sewers to Esholt, where they separated it out (because other things come down sewers besides wool fat), purified it and sold it to the cosmetic industry. Wool fat is the basis of lanolin, which is used in many facial creams."

If I understand correctly, waste matter (specifically, sheep sweat) extracted by Bradford Sewage Works has been diverted to the cosmetics industry for long decades past, demonstrating the eternal cycle from foul to fragrant and vice versa, one of the wonderful properties of nature.

"Many was the time me and my mate Ronnie Moreton, instead of getting the bus home, used to walk along the Aire Valley through Esholt's sewage works, and we would go home that way on a Friday evening, just to celebrate the end of a working week."

Ronnie Moreton was hot on science and mathematics, but not so interested in art and music. Whereas Donald Smith was a jazz aficionado. Donald Smith, Bill explains, was the son of a Methodist minister.

"He's probably still the son of a Methodist minister. He might *be* a Methodist minister."

Donald introduced Bill to jazz, almost certainly the New Orleans variety. 1947 was not too early to be modern in Minton's Playhouse, but Parker and Monk had yet to reach the Aire Valley.

21 Bill Leader, interview with the author, February 13, 2015.

22 Bill Leader, interview with the author, March 15, 2012.

23 The official title these days is Esholt Waste Water Treatment Works.

"Ronnie Moreton was hot on science and mathematics, but not so interested in art and music. Whereas Donald Smith was a jazz aficionado" (collection: Bill Leader)

"He was well-informed. He was very Methodist Christian in his outlooks, but he would not knowingly do anyone any harm. So he was a mate."

And there was a slightly older guy, Tom Durkin, who wasn't an apprentice but recently demobbed from the RAF.

"He was an Irish Catholic, from a very Irish family, locally, and he had a broad general knowledge, and a broad cultural appreciation too. When we worked at benches across from each other, filing and drilling and assembling things (because we were working by hand), I learned quite a lot from him. Because he was a progressive minded lad, although not politically committed in any way.

"It kept me out of the army. As an indentured apprentice, I didn't have to do my conscription at 18. I was let off until I'd finished my apprenticeship at 21. Later, jumping a bit, I actually got a job at Apperley Bridge Station, working for British Rail. That was quite interesting, because, as the nearest station to Esholt the sewage works, they were used to dealing with Esholt's effluent too.

"As you went from Shipley to Leeds by train, you passed a factory with a big sign which said *GRANCETA*, and they made fertiliser. I thought 'Grancreta' was a splendid name for a fertiliser. And the station master at Apperley Bridge was involved in having to ship this sewage sludge around. He organised the freight wagons out and the freight wagons back. There's a term used in transporting goods called 'demurrage'. I think it's the charge that arises in… something to do with the turnaround of… I don't know exactly; I'll have to look it up."

(From Wikipedia: "Demurrage dɪˈmʌrɪdʒ/ *noun* LAW a charge

payable to the owner of a chartered ship on failure to load or discharge the ship within the time agreed.")

"But it's a phrase that comes up. And sewage sludge was sent to all sorts of exotic places like Drogheda in Ireland.[24] So every now and again the station master would go to the booking clerk who possessed the arithmetic skills in the station, and say, 'What's the demurrage on the sludge to Drogheda?' There's poetry there."

The railway belongs to that period of Bill's life which, with the tranquility of age, he cheerfully categorises as 'pissing about'. Apprenticeship, National Service, his early trades, including the stint on the railway and the interlude as an encyclopaedia salesman all come under the broad heading, 'pissing about'.

With the end of apprenticeship in 1951, the conscription deferment ran out and Bill enrolled in the Royal Electrical and Mechanical Engineers. His experience in Hong Kong is discussed in the chapter after next. He rejoined Airedale Electric upon completion of National Service in 1953, because employers had to keep conscripts' jobs open by law. But Ronnie Moreton and Tom Durkin had moved to English Electric. Bill followed them and worked on the assembly-line. "I can't spend my life here" was his constant thought. Bill was no more suited to English Electric, "one of the more enlightened employers", than he was to Airedale.

Feeling the need to make something of his life, Bill became the chairman of the Bradford branch of the AEU Junior Workers' Committee. In this capacity he was sent as delegate to various national conferences, including a Labour League of Youth event at Butlin's holiday camp in Filey. Alas, the film *South Riding*, the most enticing attraction, stubbornly resisted all attempts at screening: "And they'd cleaned off the projector in the cinema specially."[25] (It was *Carry On Henry* when I was at Filey Butlin's, I remember.)

Tom Driberg was present.

---

24 Drogheda is 56 km north of Dublin, and historically within the Pale. The great traditional singer Donal Maguire comes from Drogheda and the town is immortalised, alright mentioned, in the chorus of 'The Pride of the Coombe', which Donal has been known to sing – *You may travel from Clare to the County Kildare / From Drogheda right back by Macroom / But where would you see a fine widow like me / Biddy Mulligan, the pride of the Coombe, me boys / Biddy Mulligan, the pride of the Coombe.*

25 Bill Leader, interview with the author, June 15, 2012.

"Of course, we knew nothing about his sexuality, but he was an immensely popular person. He had charisma. Of all the people that came down from Transport House, he was the one that everybody warmed to. The only other time I was close to him (within a foot or two), was one of the more modest places in Old Compton Street. He was there with Joan Littlewood and someone else. They'd obviously gone to meet and discuss something or other. That was fairly late on. Late on for me, that is; it was the late seventies, I suppose." (Late for Bill, and even later for Driberg, who died August 12, 1976.)

Disgruntled with life and overcome by wanderlust, Bill resigned his job at English Electric and applied for a job at British Rail. The interview took place in York, headquarters of the North Eastern Region. An eye test, mandatory for applicants, confirmed what he had suspected since taking a colour-blind test in a colour supplement a few weeks before: he was colour-blind!

"I came to the conclusion that being colour-blind meant that you couldn't pass the colour-blind test. I thought, they won't let me be an engine-driver, but unlike most kids, I didn't want to be an engine-driver. I wanted to be a sound engineer. And they wouldn't let me fly an aeroplane. OK, I can get through life without flying an aeroplane. What does it matter being colour-blind?"[26]

And so it was that Bill became a ticket-clerk at Apperley Bridge Station, and, before that, at Shipley Station, and concurrently (in the evenings), at Guiseley Station. Bill's predecessor on this line was Branwell Brontë, who, in 1841, transferred from Sowerby Bridge to Luddenden Foot (Helen Leader lives at Luddenden Foot). Branwell was sacked for not being able to add up and for arriving at work drunk at breakfast-time.

(The foregoing paragraph has undergone scrutiny by our railway adviser, Stanley Accrington. Stanley writes: "Some railwayman will pick this up as an obvious error. To the non-railway mind, Bill worked at a Yorkshire railway station near Bradford, and so did Mr Brontë, but they are not connected lines, contrary to what is stated. Apperley Bridge is north of Bradford and was on routes that emanate from Leeds or Bradford Forster Square. Luddenden Foot is on the line that goes east-west through Bradford calling at Bradford Interchange, and goes on to Halifax then Lancashire. To make the link with the drunken

26 Bill Leader, interview with the author, March 15, 2012.

clerk Brontë, which is an interesting story in itself, you would have to say something like, 'Bill became a ticket-clerk at Apperley Bridge Station to the north of Bradford… not many miles away from where notorious predecessor Branwell Brontë… at Luddenden Foot' etc. I think it's worth being more accurate. By the way, the folk world is full of potentially nitpicking railway buffs. I must also plead guilty of this charge for when I see a locomotive shown on TV purporting to be in a location it never would have been, I reach for my virtual pen and write, 'I think you'll find that Southern Region Schools class locomotives would not have been hauling the Flying Scotsman…'"[27]

Bill's last word on the subject, addressing Stanley: "To some people all railway lines are the same. There's only one railway line and it wanders all over the country. But we know differently, don't we?")

Bill's career at British Rail was scarcely more illustrious than that of Branwell Brontë. He received a blow when he discovered that free travel was confined to the North Eastern Region. He could visit York, but not London. Bill didn't think ill of the management: "I just thought it wasn't fair."

Bill next became a travelling salesman for Caxton.

"I suppose I was just generally unsettled anyway. So when I saw an ad for a rep for the Caxton Publishing Company – producers of encyclopaedias and books of instruction of all sorts, who worked in the great tradition of foot in the door book salesmen – I applied, and got the job, which should have made me suspicious straight away. I showed no aptitude at the interview."

Specifically, Bill was tasked to call on garages to persuade apprentice mechanics to invest in daunting tomes of technical instruction.

"It was a multi-volume encyclopaedia, *All You Need to Know About Being a Car Mechanic*, or *How to Mechanic Cars*, or something. I had to lug a sample copy around. And of course, you'd walk into a garage and the fellow you saw was probably the proprietor, who had no intention of letting any of his lads get into debt to the Caxton Publishing Company. So you got the bum's rush. And I couldn't deal with that. I had no idea how you sold things. It didn't come naturally to me. It was something I knew nothing about."

"How old would you be when you were a travelling encyclopaedia salesman?"

27 Stanley Accrington, email to the author, July 1, 2020.

"I was 24, I suppose. Old enough to know better. I was married. Oh, there's a thing. I'd had a girlfriend called Gloria since I was 18. In 1951 I joined the army. We decided there would be more money floating about if we got married (there was money paid to the wives of soldiers, you see). So we got married in '52 before I went off to Hong Kong.

"Fancy skipping around that one."

I had read of love, but dismissed it in my ignorance,
thinking it was an imaginary thing, like heaven.

MALACHI WHITAKER, *AND SO DID I*

V

Hardcastle Hearts

"ONCE A YEAR, AND always at the same time..." Bill furrows his brow. "And I'm trying to think when. Easter was too early, and summer was too busy. So I think it was Whitsun. Once a year, the Young Communist League camped on the borders between Lancashire and Yorkshire, at Hardcastle Crags."[1]

"We belonged to the Young Communist League," says Gloria. "They had this meeting at Hardcastle Crags. It was lovely. I think everybody met their future spouses."[2]

Hardcastle Crags, located above Hebden Bridge in West Yorkshire, is a renowned beauty spot. I have picnicked there, lolling beside the stream with dragonflies wafting past, chasing ants out of the potato salad whilst discussing how much bloodshed can be countenanced for the sake of revolution. It was then I was struck by a spooky thought: I was repeating,

1 Bill Leader, interview with the author, March 15, 2012.

2 Gloria Dallas, interview with the author, July 20, 2016.

probably word for word, the conversation young communists had on this very spot. When I ask Gloria. She says, no, they talked about something else. They talked romance, not revolution. Hardcastle Crags is the enchanted wood of *A Midsummer Night's Dream*: a space where lovers find their hearts' desire after mix-up and confusion.

"The first time I went to the YCL camp at Hardcastle Crags was, I suppose, 1947. I'd be 17. It's a beautiful spot, and we'd all gather in the evenings around the campfire and sing songs: folkie-type songs that were beginning to get very popular. The MacColl and Bert Lloyd repertoire, plus the usual student-type songs, and songs of the working class movement – all the demonstrations in the thirties against unemployment and the threat of war had lots of singing – plus the usual community-type songs you sang before the war to keep cheerful, like 'My Old Man Said Follow the Van'. 'The Quartermaster's Store', a scout song, was very popular. And then there were Soviet songs. There was something called 'The Airman's Song'. Yes, there was no shortage of songs to sing around the campfire at the joint Lancashire and Yorkshire Young Communist League camp, whenever it was held. As I say, I suspect Whitsun."

Whitsuntide was a week of holidays between Christmas, which was always cold, and summer, which was always hot.

"In 1948 I went there and met this lovely lady called Gloria."

(Whittington is her last name.)

"We walked down from Hardcastle Crags and went to Hebden Bridge Cinema on the first night."

Pressed to name the film he and Gloria watched, he offers: "It was an English film and it was in black and white. It was sort of John Millsy, and terribly patriotic, and featured submarines."

Bill ponders some more.

"It wasn't *In Which We Serve*, because I've seen that once and never intend, however often they tempt me to watch it on television, to see it again."

In Which We Serve has John Mills, Richard Attenborough and Noel Coward. Bill has a down on Noel Coward. I sympathise without knowing his precise reason. *We Dive at Dawn* (1943) is superficially similar and shares a plural personal pronoun, except John Mills is an officer rather than a squaddie, and it doesn't have Noel Coward.

"We won, I think, in the end."

Hardcastle quartet – Bill and Gloria and Louise and Alex

Gloria and Bill's dad (collection: Bill Leader)

Gloria on Ilkley Moor

In that case it must be *We Dive at Dawn*. But note how the film took four years from general release to reach that pleasant backwater, Hebden Bridge.

Gloria doesn't remember *We Dive at Dawn*. She doesn't remember going to the pictures at all. And then she drops this bombshell: "I'd met him before. We used to go to a study group in Bradford, about Engels and all those political philosophers."

"My memory is Hardcastle Crags was our first meeting," says Bill, firmly.

Music plays a part in Gloria's earliest impression of Bill.

"I know that he'd been into Hebden Bridge with friends, and they'd come back, each singing different parts of music, classical music. I wasn't with them on that occasion (I don't think I was anyway). I don't know why it made such a big impression on me. Perhaps because I didn't know enough about music to be able to walk along with someone and sing one part while they were singing another part."[3]

Bill is too gentlemanly to contradict this account, and reluctant to disabuse the notion of his advanced musicality.

"So you made a hit with something she wasn't there to see, doing something you didn't do?"

"That's the way to do it," replies Bill, with a twinkle.[4]

The question of whether Bill was capable of such a feat might be worth considering, at the risk of keeping young love waiting in the wings.

Musical education and sentimental education were linked for Bill from the start. Joyce Tetley, a girlfriend from Shipley Youth Centre, was an accomplished pianist and sang. "She had quite a nice mezzo voice," says Bill. Sometimes she would play a Beethoven sonata or Chopin nocturne and talk Bill through it. They would go to concerts in Eastbrook Hall, a large Methodist building, or perhaps see foreign films "at the excellent Civic Theatre, run by the brilliant Esme Church".[5]

---

3 Gloria Dallas, interview with the author, February 27, 2013, and all subsequent quotes by Gloria unless otherwise noted.

4 Bill Leader, interview with the author, October 15, 2014, and also for the Joyce Tetley reminiscence, below.

5 Bill Leader, by email, September 25, 2018. Actress and theatre director Esme Church became artistic director at the Bradford Civic Playhouse in 1944, founded The Northern Theatre School in Chapel Street, using the theatre's facilities to teach a generation of notable actors (Tom Bell, Billie Whitelaw, Robert Stephens, Bernard Hepton, *et al*).

Mr Pybus was responsible for Bill's formal music education. Pybus (Bill always calls him 'Pybus') taught a music appreciation class given by the Workers' Education Association, oft shortened to the WEA. These were extra-mural courses given by university lecturers (mainly alumni from Leeds University). While Bill Snr and Lou spent their Sundays attending E.P. Thompson's course on political history, Bill was in a school not far away, listening to Pybus expound on the major musical achievements of Western culture.

He might explain how the baroque orchestra was divided into two parts; the most talented musicians would perform as soloists, whilst the journeymen musicians, the greater number, would play the *ripieno* (Italian for 'stuffing' or 'padding'), in sections marked *tutti*. He illustrated with extracts from *Brandenburg Concerto no. 2*, (or some such), played on the classroom piano. He described how music modulated from one key to another. This was entirely theoretical to Bill. His parents didn't have a piano. "I recognise that things do change keys, and I can see why they do."[6]

The class studied Elgar's *Enigma Variations* and the same composer's symphonic study *Falstaff*, as well as J.S. Bach's *Violin Concerto in A Minor* and *Appalachia* by local hero Frederick Delius.

"What else? Oh, we did quite a bit of Sibelius. We took Sibelius' *Second Symphony* apart and put it together again."

Sibelius was at the high-brow end of the spectrum for many. Pybus relayed his knowledge with sensitivity and tact, and whilst he never endorsed the widely held view that Gracie Fields was the greatest singer of all time, he didn't flatly deny it.

Flash forward to May or June 1958. Bill is now a sound engineer and wandering unattended in the corridors of St Pancras Town Hall. He is about to experience a close encounter of the Pybus kind.

"In the days when St Pancras had a Town Hall – in the days when there was a St Pancras to have a town hall – I was there quite often. It was built, like all town halls were built, to house whatever happened to be required at the time, be it a symphony orchestra, or a boxing match. St Pancras Town Hall was where I first heard Stan Kelly sing 'Liverpool Lullaby'.

"I was there for some reason or other that I don't remember. Anyway, I found myself wandering around the upper floors, and I could hear this music playing. Ravel. So I thought, *ooh*! I followed it. I thought

6 Bill Leader, interview with the author, April 25, 2013.

it must be coming from the main auditorium. And then I found there was a door into the Circle of the main hall. And as I opened the door, the sound got louder. It was a wonderful, wonderful sound. And I walked down to the front of the Circle, so that I was looking into the hall. I found that I was over the top of the London Symphony Orchestra. They were actually rehearsing in the body of the hall, not on the platform, and the conductor was Pierre Monteux. He had been a conductor in Paris for many, many years. He was famous for his Ravel and Debussy. He was most famous, of course, for having conducted the orchestra for the first night of the *Rite of Spring*, which turned into the *Riot of Spring*. That was, what? 1913 or something.

"So this fellow was now an elderly gent.[7] But conductors live a long time. I think all that arm-waving is good for the heart. And there was Monteux conducting the London Symphony, at that time not in the league of world orchestras but famed for its string sound. That was what was playing: Ravel, *Daphnis and Chloe*. And there was I, right over the top, not hearing it from a distance, with the orchestra on the platform, and me on a seat in the auditorium. It was as if I was suspended over the top of the orchestra, with the hall all around. It was a wonderful experience, sonically and musically. Nearly as good as when Stan got up and sang 'Liverpool Lullaby', but on a different level."[8]

We are no closer to finding out if Bill could essay parts from light opera whilst in a state of pleasant intoxication. The tale will have to be marked 'unproven', along with several other disputed episodes of cultural life in late forties/early fifties Bradford. Did he (as Gloria suggests) collect Mahler symphonies on 78 and play them on a gramophone in his bedroom attic at 13 Park Grove? No, says Bill, the expense would have been prohibitive, *and* it would create a storage problem. Again, did he seek out performances of obscure operas because he was jaded with the popular repertoire? About half and half.

"If you'd never heard of it, I would go see it. The same with records. If you'd never heard of it, I'd buy it. You don't want to have records of stuff you can hear any old time, do you?"[9]

It was in this spirit Bill whisked Gloria off to see Vaughan Williams' *Pilgrim's Progress*, in which the great man took a bow at the curtain call,

---

7 Pierre Monteux was 83 in 1958.

8 Bill Leader, interview with the author, October 3, 2014.

9 Bill Leader, interview with the author, October 15, 2014.

and, on another occasion, the same composer's *Hugh the Drover*. With its street cries, ballad sellers, morris men and village pugilists, *Hugh the Drover* depicts an England lifted whole from broadsheet ballads. It was a touchstone for Stan Kelly; it seems he associated it with a lost love called Shelley Vaughan Williams, a relation of the composer, the author of *When Gazelles Leap*, a volume of romantic poetry. John Foreman might have modelled himself on the character of the Ballad Seller from *Hugh the Drover* ("Who will buy my ballads and songs? Three yards a penny.") What does *Hugh the Drover* mean to Bill?

"The actual performance is almost a non-existent memory. The name features, because I read a Pelican book once, and it was about English Music or something. It had all the composers of that era, the beginning of the English Renaissance. It featured in that quite heavily as a significant work."

The higher life, it seems, was never far away, in post-war Bradford.

"Bradford and Leeds were the twin capitols. Leeds was in fact less effective. The Yorkshire Symphony Orchestra was based in Leeds. It didn't last very long. The main conductor for most of the time was a bit of an alcoholic. It didn't go too well. Whereas Bradford didn't have the energy, or didn't have the ability, to generate a symphony orchestra of its own, so it contributed to the circuit of the Hallé, and the Hallé was under Barbirolli at that time, so one of its absolute peak periods.

"Bradford was strong certainly. The music wasn't bad, but what was extremely strong was the Civic Theatre, that was constantly breaking new ground. So yes, it was great to be there. If you couldn't be in London, you might as well be in Bradford."[10]

"He started off my education," says Gloria of Bill. "Although my sister and I used to listen to opera on the radio, we didn't go to anything. Heather and I used to go to concerts of the Hallé."

Heather Wallis was Gloria's best friend at school. Her father was a trade union organiser in Leeds.

"They were a lovely couple, and they had a lovely family, and Heather was the eldest one."

The curriculum at school included trade union history: Miss Mitchell didn't let her Conservative leanings interfere with her work, happily. The two friends sold the *Daily Worker* on the street. But now Heather was exasperated because Bill was the dominant subject of

10 Bill Leader, interview with the author, July 22, 2020.

Women of the Bradford Young Communist League, with Heather Wallis and Gloria Whittington, front, and Mrs Wallis, first in the second row (collection: Bill Leader)

conversation on their cycling holiday of France, undertaken shortly after the YCL camp. Heather couldn't see it. She had been paired off with one of Bill's friends.

"I remember that Heather and I didn't want to spend our time in a pub. We insisted on going to a coffee bar instead of going to a pub. And they accepted, which was a great surprise to me. We didn't realise what enormous power we had over men."

Bill believes Heather saved Gloria's sanity.

"Because Heather came from a very stable family, a happy family background, and Gloria's was anything but stable. She wouldn't have believed a steady relationship was possible if she didn't know Heather. We proved, of course, that it *was* difficult to have a steady relationship. But we did our best."

"Heather's mother and father were in the Communist Party," says Gloria. "So were Bill's. So when I was telling them about him, they said, 'Oh he's that nice-looking boy of the Leaders.' Bill has very bright blue eyes, and his parents…" Like the artist she is, Gloria captures the love between Bill Snr and Lou in a single image. "I walked into the room where they were sitting, one on each side of the fireplace, and they'd both got white hair, and there were these two pairs of bright blue eyes."

"This is Gloria and one of her fellow art students and two other people whose names I don't know. She did a holiday job at Butlin's, Filey. These are the people who worked with her, long-term Butlin workers, whereas the two bookends, Gloria and the girl at the other end, were Bradford Art College students"

Glam Gloria
(collection: Bill Leader)

The suitors were plagued by junior family members.

"Gloria's brother, Tony, and me used to annoy Bill and Gloria if we possibly could," remembers Bill's cousin Jo. "We used to go fishing on Sunday mornings in the canal, and then Aunt Lou would say, 'You can't go in the front-room because Bill and Gloria are in there.' Well that was a stupid thing to say to an eight-year-old and an eleven-year-old, wasn't it? We went into the front-room, didn't we! Gloria hated being called 'Glo', so Tony used to call her 'our Glo'. And we'd creep in behind the settee and try and disturb them and annoy them. Tony died a long, long time ago."[11]

Gloria was at school and Bill serving his apprenticeship when they met, but soon she moved to Bradford Regional College of Art, part-financing her studies with holiday work at Butlin's holiday camp in Filey. It was a vocational art college, with an emphasis on textiles, graphics and 3-D. David Hockney attended the school in 1953, the year Gloria left, and, although their paths didn't cross, she followed his career with interest. Gloria opted for the painting course, but this didn't preclude an interest in songs and singing. There was a small atelier called the Little End Room in Bradford College of Art, which had space for only a few easels. It was used by students who wanted to put in extra brushwork after-hours.

"And I used to sing while we were painting. Nobody objected, so I suppose it must have been alright."

Sometimes the Little End Room rang to Kate Haley's singing. Kate was a life-class model of Irish heritage.

"It's funny to say someone's Irish when they've lived all their life in Batley, but Kate was Irish. She had the most fantastic repertoire, all picked up from these Saturday evening sessions in Huddersfield."

The singing would carry on at Kate's parents' house after the pub shut.

"She taught me 'I Love My Miner Lad'. Stefan Grossman used to say that you should try and do a song exactly as you heard it, but you won't succeed. Long before I heard him say that I got this song from Kate, and I sang it exactly as she sang it. Bill's mum said, 'You sound drunk!'"

'I Love My Miner Lad' is a different song from Roud 2599, 'My Bonny Miner Lad',[12] but the sentiment – unconcealed lust for a handsome miner – is the same.

---

11 Jo Pye, interview with the author, December 11, 2013.

12 As sung by Shirley Collins on her second LP, *False True Lovers*, and, later, on her Collector EP *English Songs Vol.2*.

"Then somebody from Leeds University came and brought a tape recorder and recorded her singing."

This was Tony Green, who went on to contribute to Leeds University's Song Society magazine *Abe's Folk Music*, edited by Bob Pegg, and became a professor at the same institution. Kate married her bonny miner lad.

"She married the same miner lad she sang about in the song," says Gloria. "One day he [Tony Green] came along, knocked on the door, and this great big tall man answered and said, 'Are you the one that's been bothering my wife? *Well don't come again!*' That was the end of that. I've met her since, and she only sings funny songs at family get-togethers now. I sang her the song she taught me. She didn't remember it."[13]

While Bill "was busy pissing about not building any career at all", Gloria was studying for a Diploma in Art Teaching at Leeds College of Art. Because her dissertation had to be about something other than her specialist subject (good old General Studies!), Gloria chose to write about folk song collectors. She sweated over the project for long hours in the reference section of Leeds Central Library, soaking up in the high Victorian atmosphere, which was appropriate too, because one of her subjects, Frank Kidson, witnessed the construction of the buildings – town hall, library, courts – on a field near where he lived as a child.

"There were lots of wonderful things there," says Gloria, "like *Blackletter Ballads*, a book written by Frank Kidson's daughter[14]

---

13 'I Love My Miner Lad' appears, as 'Miner Lad', on the 1972 Jacqui and Bridie LP, *Next Time Round*, Galliard GAL 4091, with the credit 'Trad. arr. Gloria Leader'. Jacqui and Bridie's source was likely *Sing*, vol. 2, no. 4 (Oct.-Nov. 1955), p. 52. John Hasted, *Sing*'s music editor, transcribed the song from Gloria's singing. It has an informative note: "Taken down in Yorkshire from the singing of KATE HALEY and contributed by Gloria Leader. The tune appears to be a variant of 'Drumdelgie' and the words have a lot in common with 'Six Jolly Wee Miners'. See A.L. Lloyd, *Come All Ye Bold Miners*, Lawrence and Wishart, London, 1953."

14 In fact Ethel Kidson was Frank Kidson's niece, who he adopted as his daughter, and who subsequently became a close collaborator and continued his work upon his death in 1926. Pete Coe is a leading Kidson authority, and has a CD and touring show called *The Search For Five Finger Frank*, cf. http://fivefingerfrank.co.uk. This was the source of the titbit about Kidson and Leeds Central Library.

Bill and Gloria wed one day in 1952,
prior to Bill's posting to Hong Kong

about her father, which is just a notebook, like a school exercise-book. I used that. And then one day I went to find it and they'd filed it in the wrong place. Well, if something is filed in the wrong place in a library like that, it's never going to be found again. But it was a wonderful thing she'd written."

The folk revival was in its infancy and sources for Gloria's dissertation were scarce. Ewan MacColl had yet to cross the divide from theatre to folk, and A.L. Lloyd's book *Come All Ye Bold Miners* was in preparation. *As I Roved Out*, on the BBC Light Programme on Sunday morning, was the only outlet for traditional music in the media. There were murmurs of a mythical figure from the States called Woody Guthrie, but he was mainly a rumour.

"*As I Roved Out*, with Peter Kennedy. It was very good. But I do call Woody Guthrie 'Willie Guthrie', so I wasn't familiar with everything," says Gloria, ruefully.

Bill and Gloria were married one day in 1952, prior to Bill's posting to Hong Kong. Gloria wore a smart New Look coat. "New Look was fairly new in then," says Bill.[15] The small wedding party comprised

15 Bill Leader, interview with the author, June 7, 2019, and for all comments about the wedding.

Bill Snr and Lou, Gloria's father, Llewelyn, who was always 'Dick' because his last name was Whittington, and Gloria's mother Edith. Also present were Gloria's sister, Pauline, and young brother, Tony ("who died very young, and I don't know why"), and Mr and Mrs Wright. Willie Wright was the secretary of the local district of the Amalgamated Engineering Union and best man. "In fact, he paid for the certificate out of his own pocket, which was seven shillings and sixpence in those days."

It was a registry office ceremony without fuss or flowers. Gloria felt apprehensive. She worried that the reason for the marriage – that the army paid an extra allowance to wives of servicemen – wasn't very romantic, and she also faced the prospect of an immediate and prolonged parting from her new husband. At least she would be able to pursue her studies without interruption with Bill away.

"He was in the army. Everybody had to go in the army. Conscription, yes. Just for a certain length of time…"

Here is a picture of four servicemen in a railway carriage. Bill (for it is he) exudes mellow contentment – he appears to be horizontal – and all are merry to varying degree. The inscription on the back is revealing.

"First-class carriage on the troop train."

"First-class? I don't think the powers that be lavished such luxury on working-class conscripts. Is that what it says?"

I read aloud, "'This is us in our first-class carriage on the troop train… Personnel are, left to right, me, Lance Corporal Maclashan…'"

"He was a Glasgow heavy lad. You wouldn't want to meet him on a dark night if he wasn't feeling pleasant."

"'Craftsman Green (Bradford)…'"

"Was he from Bradford?"

"'And Craftsman Leif.'"

"I don't remember their names. I remember Lance Corporal Maclashan, or whatever his name was. That's an embarkation train heading off for Hong Kong."[16]

Felix Eaton has picked me up at Honley Station and briefs me as we park outside his mother's house. Louise's memory is not what it was, he warns. The business about Bill's uniform is a worrying example of her wool-gathering. It's impossible that Louise ever saw Bill in uniform (as she confided to me on the night of the Gold Badge Award). The dates don't fit. She knew him through Alex, Felix's father, and Bill had left the army before the two knew each other. As I listen, I find I'm distracted by the mellifluousness of Felix's voice. He has been identified as his father's son just by speaking. The matter of the uniform comes up as soon as we cross the threshold.

"When you've been in the army are you periodically called up to have a refresher or anything like that?" asks Louise, anxiously.

"Not that I know of," says Felix. "Once you've done your National Service you've served your debt to your country. It's just an incorrect memory."

Louise is beginning to doubt her own senses: "You can make things up, can't you?"

"It's possible. You'll have to ask Bill."

I ask Bill. It seems the bulk of his service fell in the gap between the reigns of George VI and Elizabeth II.

"I'd gone in the Army under King George VI, god bless him, but he died just before I went off to Hong Kong, and, on my return journey on a troop ship back home, we stopped at Colombo Harbour (Ceylon as it

16 Bill Leader, interview with the author, February 14, 2020.

then was) and listened to the coronation of Elizabeth II."[17]

George VI died on February 6, 1952, and the coronation of Elizabeth II occurred on June 2, 1953. In 1953 Louise was not yet a dancer at Idle and Thackley Amateur Dramatics. She progressed to Persephone, a lady's clog-dancing team, and, in 1995, organised the first Day of Dance, 'a celebration of dance in all its variety', at Victoria Hall in Saltaire. It was so successful it became an annual fixture. Upon retiring as a dancer, she took up the tenor horn. It sits on the chair beside a music stand in the living-room as we speak.

Louise rings Gloria and, receiving an automated reply, leaves a message: when did Bill leave the army?

The conversation turns to the courtship of Louise and Alex Eaton. It started with a French class.

"I went to French evening classes taken by Alan Bennett." (This Alan Bennett was not the Leeds playwright and national treasure, but a French teacher at Carlton Grammar School, where Alex also taught.) "And Alan couldn't go one evening, and said, 'Would you take the evening class for me, Alex?' And so Alex took the class, but just the once.

"Between them, they were organising a trip to take the fifth-form boys to Paris for a week. Alex said to Alan, 'Why don't you suggest if any of the evening class would like to come?' I put my hand up straight away, as did this other girl, but then she chickened out. She was too shy or frightened to go. Cicely used to sit on the bus with me when I went to work (she got on, oh, about half a mile before me). I said to her, 'Would you like to go to Paris?' She was up for it. She wasn't shy. And yet she wasn't pushy either. She was very modest. She was an aspiring soprano and the soloist in Idle and Thackley Amateur Dramatics. I joined as a chorus-girl and dancer. So we met up with Alex and Alan Bennett and we were to go on the first of April 1955. April Fool's Day. But it wasn't foolish for me. It was wonderful. We fell in love there, straight away.

"I was this shy lass… Well I wasn't shy; I've always been outgoing. I've always had to push myself to overcome my shyness because, of course, I'm an extrovert at heart. It sounds a contradiction, doesn't it? And, oh, I used to quiver. But I couldn't keep quiet.

"So we set off on this trip. Alan Bennett is a very staid chap. I wasn't

17 Bill Leader, interview with the author, March 15, 2012.

attracted to him at all, and Cicely was very attracted to Alex as well. Any woman would be. Alex was surprised at this. So there we were: I sat by the window (we got in first), and Cicely sat opposite me on the other side of the table. There's a tunnel, quite a long, dark tunnel as you leave Bradford Station. It seemed pretty early in the journey to me, but perhaps it was far later, and I misremember."

Felix, who knows the story well, professes uncertainty about the precise location of the tunnel.

"We went in this tunnel, and all the lights went out. Usually they keep them on, but this time the lights went out. There was this kiss, from Alex." Louise laughs. "What you might consider a full plonker. Well, I got shivers right up and down, and as soon as the lights came on, we just sat there. I don't know whether I'd got red cheeks or not. He told me he wanted to firmly plant his possession. He was frightened of Cicely. She was a very heavily made-up woman. I didn't use any make-up at all. Cicely was heavily made-up, and he hated it. He wanted this natural looking lass, if he was to be landed with anybody. And clearly that was their idea, to lighten the care of the boys. We spent time with the boys. We did our duty."

Louise goes to fetch a photograph album. Here, Alex and Louise are walking by a fence having just visited the Sacré-Cœur Basilica in Montmartre. Of her beauty, Louise now says, "I didn't feel I was beautiful at all. I always felt I was very plain. It's only when I look now at photographs. I think, 'Gosh, I was quite nice looking, wasn't I?' I didn't feel so at all."[18]

They make a good match. Alex, strikingly handsome with his head thrown back in the exaltation of song; now it's Louise's turn to play guitar, and her lovely smile becomes tinged with uncertainty; here, Bill, Alex, Louise and an unknown friend stand in front of a hut in a sylvan setting. From the evidence of the photo-album the *annus mirabilis* of 1955 was spent mostly at Hardcastle Crags or Wortley Hall, with the odd excursion to Paris. Bill is sticking his tongue out, the unknown friend looks mildly puzzled and Alex and Louise are in a bubble of bliss. Pausing only to swap photographers (Louise takes over from the unknown friend's companion), Alex is seen sucking in a lungful of smoke, as Bill peers inscrutably ahead, cigarette in hand. "Lots of cigarettes in those days," remarks Louise, reading my mind. In

18 Louise Eaton, interview with the author, February 25, 2015.

another, Alex snatches a moment from camping to romp in the grass with Louise. The Young Communist League can wait.

And here is a sloped lawn in front of a familiar stately home occupied by, from left to right, a coquettish Gloria, a serious Bill, a smiling/grimacing Alex, and a gentleman, older than the others, sprawled in a smart suit. And in the last photograph (or the last I can be bothered to describe), John Hasted is demonstrating guitar to a party who have their heads down perusing something in their hand-outs. They are all female, except for one man. Is it Alex? The ornate stone urn confirms that we're back at Wortley Hall. It can be nowhere else.

The phone rings. Gloria is returning Louise's call.

"Hello Gloria!… I'm fine, thank you. I've got somebody here researching the life of Bill. You've met him… Yes, he said he'd been to see you. I was trying to recall when Bill first went into the army, but it was before my time apparently… 1951… You were 20… So did you know Bill when he was conscripted?… Oh, you were married! Ha ha!… Obviously Mick [*sic*] knows all this because he's talked to you, but I didn't… That's right, Hardcastle Crags. Yes there's a picture of us, of me and Alex there. And there's a picture of the four of us. We're sitting on a slope… Is that Alan Bush? Can you remember the picture, Gloria? There's an older man sitting in front of us. Wasn't Alan Bush a conductor?… A conductor and a composer. That's right. I think it's Alan Bush… Wortley Hall, this is. Sorry. Apologies Gloria, Wortley Hall I'm talking about. Yes, and that photograph we've got of you, looking all coy, looking over your shoulder. Ha ha! And there's Bill on the other side of Alex and clearly I'm taking the picture. I always thought the bald-headed chap was Alan Bush… Alex was a bit bald-headed, and even Bill had a high foreline, as young as he was. This was nineteen…fifty…ah…"

The uniform did figure in my conversation with Gloria. Bill was so careless of the honour of the regiment, that, home on leave, he stood on the toilet seat in uniform to fix the cistern of the lavatory. Gloria was so charmed she made a sketch on the spot, but it has since been lost. As we spoke, my eye rested on a line drawing of Evan Parker on the kitchen wall.

Too much is being made of the uniform, I decide. At worst, it's a minor continuity issue. Sure, I would like to nail the truth (ah, but what is truth?), but other things are preoccupying me. Like, did Bill

Alex and Louise, just having visited the Sacre Coeur, Montmartre; camping and frolicking (collection: Louise Eaton)

Alex in song

Alex, Bill and two friends at Hardcastle Crags (collection: Louise Eaton)

Louise with guitar

On a sloped lawn at Wortley Hall with Gloria, Arnold Goldsbrough, Alex and Bill

John Hasted takes a guitar class, Wortley Hall (collection: Louise Eaton)

Camping and frolicking (collection: Louise Eaton)

record Frank Harte's *Dublin Street Songs* in Dublin or London? Did Billy Pigg really tell Bill that recording *The Border Minstrel* might have to be postponed until after the harvest, when, by my reckoning, Billy had been dead for three years at the time? These kinds of things keep a biographer awake at night. For Felix, however, Bill and his uniform is a test of his mother's competence. Bill is the only person alive who can settle the matter. This is what Bill says:

"I limited myself to two years National Service, which was the minimum you could do, but they promised to make you into a non-commissioned officer if you were good enough to give them an extra year, which I didn't. But for some years afterwards, it was the custom and practice, in fact the requirement, to attend your local Territorial Army drill hall frequently and do a sort of refresher course once a year. I don't think I saw the inside of the Shipley drill hall very often. I went to two camps, as required. I came out in '53, so in '54 or '55 I could have got on a train and gone somewhere down in army country, Aldershot, Salisbury Plain, with my uniform on. That's the only opportunity she might possibly have had. But I don't recall it happening."[19]

On consideration, Bill thinks the man on the lawn is Arnold Goldsbrough, the founder of the English Consort Orchestra, and not Alan Bush.

Gloria does her best with Louise's enquiry and rings off, and Louise resumes browsing through the photograph album.

"I think that's Frank!" she says of the unknown friend. "This is the hut where you could go in if it was raining. Sometimes people slept overnight there. I was camping on my own, with my own little tent, and so was Alex, and they said, 'There's that hut you know. You and Alex, you can go up there and you can be on your own.' And I said, 'No, no, I'm happy where I am.' I don't know who suggested it, whether it was Bill or not. It might have been Harry."

---

19 Bill Leader, interview with the author, February 27, 2015. He makes the same point in an email, April 6, 2020: "It comes to me that after lads finished their period of National Service, they had to spend time in the Territorial Army (which required attending the occasional drill night in uniform, and an annual week-long camp also in uniform). Louise might have come across me on one of these occasions."

"Harry Wright?" asks Felix.

"No, it was another Harry who lived in Leeds, and his wife. They could see that we were in love and we should be really cuddling up together in the hut, but no. Both of us felt it was inappropriate."

Louise dates the famous YCL camp as August 1955, almost certainly in error. Bill was already in London by then.

"Alex had his guitar. We sang YCL songs, peace songs. Alex probably sang his cowboy songs. He started a YCL choir in Leeds. That's where we met when we came back from Paris. There must have been about twelve in the choir, all Young Communists, and he was teaching them to sing left-wing songs. When there was a meeting, we'd open the evening by singing some songs.

"Eric Sandmeier," Felix prompts. "Eric is the reason you got married."

"Yes!"

Felix tells the story. "My dad was on the way down to London with Eric and confided to him that he didn't know what to do – he was a divorcee, he was older – about this relationship with a young slip of a thing. He wasn't sure whether to offer to marry her or not. And Eric said to him, '*Grasp the rose firmly. If there's a thorn, you'll soon find it.*'

"And that's what caused dad to propose."

"It did!" affirms Louise, delightedly.

He recollects the day when he durst scarcely walk the streets.
He can tell how he was hooted, pelted and spurned and people told
him he might be thankful if he was not burned alive some night,
along with an effigy of Tom Paine.

E. SLOANE, *ESSAYS, TALES AND SKETCHES* (1849)

VI

Ye Suspects of England

AMONG JO'S ARCHIVE IS a photograph of a youthful Bill in a march, carrying a placard that is impossible to read because of the angle of the shot. From the evidence of the surrounding slogans, it seems to be an all-purpose protest, with issues ranging from workers' rights to ban the bomb.

(In a film from the USA, yet to go into production, a girl at a small-town hop wonders about the emblem on the leather jackets of the gatecrashing bikers. 'BRMC' stand for 'Black Rebels' Motorcycle Club', she is told.

"Gee, well what are you rebelling against?" she asks, impressed.

Marlon Brando – for it is he – curls his lip with studied cool:

"What have you got?")

Jo gamely deciphers the messages on display.

"*AEU*, that's the Amalgamated Engineering Union, *and something District Committee Send something* (comradely? fraternal?) *Greetings to All Organised Workers*. The older man marching out at the front with a sign I can't read is Uncle Willie."

Willie Wright was the best man at Bill and Gloria's wedding and paid the 7/6 cost of the marriage certificate out of his own pocket.

We Want the 40 Hour Week. No Arms for Nazis.

"Is this the rearmament of Western Germany? What else would make them say *No Arms for Nazis*?[1] Wait a minute! That's Bill's mate."

The figure in the beret just in front of Bill is a *No Arms for Nazis* man, although almost certainly the placard was randomly allotted.

"I inveigled Ronnie Moreton into marching," says Bill. "He was a fellow apprentice at Apperley Bridge and had no interest in things political. He was quite a boffin. He ended up in English Electric in Stafford. He's quite likely to have done something noteworthy technically or scientifically. I knew him for three years. I went in the army for two years and then I knew him for at least another year afterwards, but lost track of him. I just feel he might have left his mark somewhere."[2]

Everybody is in a raincoat and it looks drab. That dates it as the early fifties. Not the drabness, drabness being perpetual, but duffle-coats would outnumber raincoats at a later date.

I share this image to those in the know, and Louise Eaton offers a dancer's insight.

"Look at their stride. They're all definitely in time. There must have been a band for them to walk in unison like that. All of them. Look!"[3]

Bill identifies the occasion as a May Day march: "Yes, I think so, with all those elaborate banners."

The workers' holiday that fell on May 1 was always honoured in the Leader household. Workers' achievements had been too hard-earned not to be celebrated.

"Tantalisingly, we can't see the banner you're holding."

"No, you can't get me for it."

"Louise pointed out that there must be a band playing, the way you're all in step."

---

1 Jo Pye, interview with the author, December 11, 2013. From John Hasted's 'Talking Rearmament' – " Pack their swastikas in a trunk in the attic / Got to run the Wehrmacht all democratic / Hitler and Goebbels are dead and gone / And eight brand new ministers sitting at Bonn…" *Sing*, vol. 1 no. 1 (May-June 1954), p.6.

2 Bill Leader, interview with the author, February 14, 2020.

3 Louise Eaton, interview with the author, February 25, 2015.

May Day, Bradford, 1951, with Willie Wright, front, Ronnie Moreton in beret, and Bill just behind Ronnie (collection: Bill Leader)

"I can't remember there being a band. I think you fall into step when you're all walking like that."

And the date?

"Well Ronnie Moreton and I were apprentices together, up until my going into the army..."

A senior moment can strike at any time.

"1971?"

"It doesn't sound right."

"1951? 1941? 1951."

"1951."

"1951. The years go past, you know."

Anyone devoted to workers' rights or committed to the cause of peace fell was under suspicion straight away. The Leaders ran foul on both counts.

"When I was quite young, I remember my father cycled from Barking for four miles in the rain, with a cape, carrying an old Gestetner duplicator on the handlebars of his bike. There was some activity going on. It was either something to do with the Peace Council, which flourished in Barking, or a strike at Ford's. I think it was probably a strike at Ford's. As he cycled as best he could with his load, from where he'd picked it up to where he was taking it, a police car followed him all the way along. He actually stopped and challenged them. He asked why was he being followed? I don't know what kind of answer they came up with, but they were there to make it clear that they were there. My dad wasn't involved in subversive or secret activity, quite the opposite.

"I was brought up to assume that they were there listening and taking what notes they wanted to."[4]

Then, over a decade later, when Maggie was staying at Shipley and daughter Jo was only a child, some trouble was brewing at Park Grove. Jo is unclear about the details; she only got half the story from Maggie.

"Aunt Lou said to Uncle Bill, 'Well send Maggie. Maggie can do it. She's not a Party member or anything.' And Uncle Bill said, 'She lives in this house. Whatever we do, affects her.' And that was the first time I really understood that you don't actually have to do anything: it's all guilt by association."[5]

For a while Bill Snr and Lou attended the class on political history by historian and peace activist E.P. Thompson, when Thompson lived in Siddal, Halifax, and taught at the Workers' Education Association. Thompson nails the conditions Bill Snr daily endured in *The Making of the English Working Class*, his most famous work:

> The classic exploitative relationship of the Industrial Revolution is depersonalised, in the sense that no lingering obligations of mutuality – of paternalism or deference, or the interests of 'the Trade' are admitted. There is no whisper of the 'just' price, or of a wage justified in relation to social or moral sanctions, as opposed to the operation of free market forces. Antagonism is accepted as intrinsic to the relations of production. Managerial or supervisory functions demand the repression of all attributes except those which further the expropriation of the maximum surplus value from labour. This is the political economy which Marx anatomised in *Das Kapital*. The worker has become an 'instrument', or an entry among other items of cost.[6]

MI5 carried expansive files on Ewan MacColl and Bert Lloyd. The intelligence agency, in a 1932 report, identified MacColl as "a communist with very extreme views". He needed "special attention",

---

4 Bill Leader, interview with the author, April 17, 2015.

5 Jo Pye, interview with the author, December 11, 2012.

6 E.P. Thompson, *The Making of the Working Class* (Victor Gollancz, 1963; Pelican 1978 edition), p. 222.

whatever that means. The report acknowledged his "exceptional ability as a singer and musical organiser" but seemed to think this was a bad thing.[7] Bill's MI5 file would make interesting reading. Presumably it sat on some dusty shelf in Mayfair (Leconfield House was the headquarters of MI5 from 1945 to 1976) and moved to Millbank in 1994. After all, Bill was a card-carrying communist, a prominent member of the Workers' Music Association, that well-known communist front, and consorted with known subversives like Ewan MacColl and A.L. Lloyd. Moreover, he had tea with Eric Hobsbawm, the Marxist historian.

"When I first moved down to London, I spent an evening in the company of Eric Hobsbawm. I don't think I met up with him after that, although he lived in Gordon Mansions, which is a block of mansion flats where various people I knew lived."[8] (That other scourge of the establishment, Karl Dallas, had a flat in Gordon Mansions, which was also the home of his ex, Gloria, and their son, Tom.)

Hobsbawm wrote about jazz under the *nom-de-guerre* Francis Newton (a nod to trumpet-player Frankie Newton, best known for his association with Billie Holiday), and gave record recitals at Conway Hall about jazz. MI5 sent an undercover agent to the meetings. He was a punctilious agent, and filed a detailed report:

> [Hobsbawm] outlined the beginnings of jazz in the New Orleans district, and its increase in popularity both in Britain and the United States of America... Jazz was, he said, mainly the music of the workers, and it had prospered in the hands of left-wingers, who were opposed to the commercialism of the music. He interspersed his speech with recordings of jazz in the traditional and modern idioms. The meeting, which commenced at 7.40pm and terminated at 9.35pm, was orderly throughout, and was attended by an audience of 27 persons.[9]

Baron Hennessy of Nympsfield needs no persuading about the real and present danger posed by the enemies of the state.

---

7 *Ruth Ewan's Did You Kiss The Foot That Kicked You* catalogue (Aldgate Press, 2006), p. 7.

8 Bill Leader, interview with the author, June 15, 2012.

9 Personal File 211764 Serial 238A Special Branch Report, cit. *The Hobsbawm Files*, BBC R4. The MI5 files on Eric Hobsbawm amount to hundreds of pages and were declassified two years after the historian's death in 2012. By coincidence, Hobsbawm was then living in Nassington Road, home to Topic Records, 1960-73.

"In a nasty world, you do need forewarning. My god you did in the Cold War. I mean, you look at the Joint Intelligence Committee watch lists, the Red and the Amber, to see if the Soviet Bloc might be cranking up for a war without much notice, or a longer term build-up to a war. My god, that mattered. And I knew it mattered. I always took it terribly seriously, because we were immensely vulnerable…

"At the core of the definition of 'subversive' in the Cold War, were those who might seek to overthrow Parliamentary democracy in the United Kingdom. And the great worry was, of course, because of the great Cold War confrontation, that they had a branch office in King Street, called the Communist Party of Great Britain. And quite a large party in the early post-war years, somewhere hovering around 40,000 for a while, having peaked in World War Two. And the real worry was: was this a legitimate political party, or was it an instrument of a power, who, with its allies in the Soviet Bloc, might, in certain circumstances, do catastrophic harm to the UK, if the Cold War balance shifted or changed to a bunch of nasties in Moscow?"[10]

Bill once attended a meeting hosted by the Music Group of the CPGB in King Street, where Alan Bush was speaker. There was a lively exchange about Stravinsky and his revisionist tendencies. The question was largely academic because the music of dissent had shifted from classical to folk. The right-wing press were swift to sound the alert:

> *A new teenage craze is sweeping Britain – folk music. Bearded, duffle-coated youngsters squat on the floor of cellar clubs listening to folk songs telling of love, of death, of oppression.*
>
> *There are more than 200 of these clubs in Britain, with 250,000 members. More clubs open every week.*
>
> *But this boom has some people very worried. For many of the movement's big names – singers, agents or record sellers – are either Communists or they hold extreme left-wing views.*
>
> *And it is feared that, with folk music attracting more and more young people, there is a danger of their being wooed by Red propaganda.*[11]

Eric Winter's little magazine, *Sing*, was just such an instrument of propaganda. It gave singers songs to sing with the aim of bolstering

10 Peter Hennessy, Baron Nympsfield, cit. *The Hobsbawm Files.*

11 'Just how INNOCENT are signs like this?' by Peter Bishop, *People*, November 17, 1963. Cited *Did You Kiss The Foot That Kicked You.*

the peace movement. Eric's widow, Audrey, recollects: "I do know that the Special Branch sat in a car outside our house. I went to offer them a cup of tea, so that was how I knew that they were there."[12]

Here is Audrey remembering, falteringly, the Save the Rosenbergs campaign, the reason for the Special Branch interest. A contemporary report in the *Manchester Guardian* relates that the Save the Rosenbergs Committee appealed directly to Winston Churchill.

"I didn't go to the vigil. I remember it and I certainly stayed up, but I stayed at home with the young kids. Sue was only a baby. I know it was a very busy time and Eric was terribly busy. I imagine the Committee did things. I shouldn't think they would bother to write to Churchill. They would regard it as useless. They were all Party members. I'm sure they would have paraded around and sung songs and done all the things that people did, but I have absolutely no real genuine recollection. At the time of course, we were convinced that the Rosenbergs were innocent, and also we were so against capital punishment. No spies in peacetime had ever been condemned to death. Wartime, yes, but not in peacetime, and so the fight was focussed around that. I'd been anti-Stalin since the autumn of '51, before Stalin died. So I wasn't terribly involved with Party things, although my husband was. I simply have no idea of the details now."

Audrey's apostasy placed Eric in the invidious position of having to choose between Party and family.

"The problem was, that although Eric and I were both in the Communist Party, and he was the branch secretary of the West Hampstead branch, in the autumn of 1951 I had a moment on the road to wherever St Paul was [Damascus], and I decided that Stalin was nothing but a tyrant, a bully and a horrible bloody awful dictator. And I said to Eric, 'I'm very sorry, but this is how I feel about it. I'm not against Marxism, and I'm not against the basic concepts of the socialist society, but I'm definitely against this idea of a nice benevolent Uncle Joe. And I went to the branch meeting and told them so. And that really put the cat among the pigeons."

"In Moscow you would have been shot."

---

12 Audrey Winter, interview with the author, April 16, 2018. This is a phone call, but it blurs with my proper interview with Audrey, on September 6, 2012, in the extract below.

"I expect so, yes. How fortunate I wasn't in Moscow.

"King Street followed the party line from Moscow, and the party line said that if your partner reneged, like I had, then you left her. I had, we had, one small child, and another one on the way, so there was no way Eric was going to leave me. It was a real impasse. Eric then went to see Harry Pollitt. They were both old friends."

Lancashire-born Harry Pollitt was the General Secretary of the Communist Party from 1929 to 1956. It was not an uninterrupted tenure – he stepped down when Stalin and Hitler signed a non-aggression pact in August 1939, and neutrality became the party line, compromising Pollitt and other committed anti-fascists. Pollitt was reinstated when Hitler invaded the Soviet Union in June 1941 and total war became inevitable. He would not be led or driven and backed the Republican struggle in Spain in pamphlets like *Save Spain From Fascism*. When it became clear that King Street was being bugged by Special Branch, Pollitt recruited John Hasted to design an electronic kit to test the phone lines.[13] After the Hungarian Uprising of 1956, he resigned from the CPGB, along with several thousand other Party members.[14]

Clearly a man of integrity, did he really deserve expulsion from heaven and instatement as the People's Commissar for Hell, the scenario of 'The Ballad of Harry Pollitt', as sung by The Limeliters? Actually, the song predates its subject's demise by three decades and is more affectionate than not. Elin Williams wrote it for a lark at Oxford in 1935. It became "one of the four best known songs in the Labour movement",[15] and there is no evidence to suggest that The Limeliters were anything other than urban left-leaning folkniks, albeit with crewcuts and a well-developed sense of fun.

A meeting was hastily arranged between Eric and Harry.

"He went to see him at home, not at the office, and Harry Pollitt, who was a very sensible chap, said never mind the Party line. He told Eric to go and do whatever his talents let him, and work for peace and democracy and those things. And that's what happened. Eric stayed in the Party. He paid his dues. But he didn't get anything through the Party. He was well out of it really. And gradually it fizzled out. I don't remember when we stopped paying dues, but we did eventually. He

---

13 See Dave Arthur, *Bert* (Pluto Press, 2012), p. 278.

14 https://spartacus-educational.com/TUpollitt.htm.

15 'Author of a Legend', *Sing*, vol. 3, no. 4 (1956), p. 56.

then concentrated on folk music, on *Sing* and so on. He was doing his ordinary job. He was a librarian, and then he went into journalism. He said, 'Right, I'll spend my time on folk music…'"

Did Audrey manage to convince her friends in the West Hampstead branch that Stalin was a horrible dictator? She didn't even convince Eric. Tellingly, MacColl's 'Ballad of Stalin' ("*Joe Stalin was a mighty man, a mighty man was he…*") was judged the Best Song of 1954 in *Sing*, three years after Audrey's apostasy. The subject was a sore point between the Winters, and Eric tended to avoid the subject in discussions with his wife. "He knew better than to argue with me, put it that way," says Audrey. Khrushchev's Secret Speech to the 20th Party Congress, made in February 1956, vindicated her stance completely, but it remained the minority view in CPGB circles. The Party faithful (Bill included) dismissed the denunciation as propaganda. Harry Pollitt, pointing to Stalin's portrait on his living room wall, declared, "He's staying there as long as I'm alive."[16]

Bill's discontent manifested itself with a daub on a wall in the centre of Shipley, and he was caught red-handed with tell-tale white paint.

"There was two of us – there was another lad who lived in Shipley – and we decided that what the world needed more than anything else was a slogan on a wall in Saltaire Road, down by The Junction, a well-known pub. So we got ourselves some whitewash and a large-sized brush and picked a night we thought would be useful for writing up the slogan, which was going to be *BRING THE LADS HOME FROM KOREA*. Yes, that was it.

"I don't know what time of night it was. It was late for us, but it probably wasn't late for anybody else. We were young and obedient kids. There wasn't much in the way of traffic on Saltaire Road. It was a time when there wasn't much traffic around. People hadn't got cars yet. They were still recovering from the war.

"So we started putting up the slogan. We got as far as *BRING THE LADS HOME FR*, or *FRO* I think it was, when we heard a car coming. We got out of the way so we couldn't be seen actually doing the dastardly deed, to a convenient gent's convenience that was right opposite. As luck would have it, as they say, the car that was coming down was a police car. So it stopped, when it saw *BRING THE LADS HOME FRO*. They looked at it. And then they started looking for us.

16 https://spartacus-educational.com/TUpollitt.htm.

"We decided the best thing would be for us to put our hands in our pockets and whistle nonchalantly and walk up the road. Which we did."

Bill and his accomplice were apprehended.

"'Hang on a minute. What do you know about this?' Of course, we didn't know a thing about it. Somehow they didn't believe us, so they took us to the local cop shop, which was only about a hundred yards away, and they questioned us, one at a time, using the time-honoured method of, 'Your mate's confessed. You might as well tell all.' My friend believed it, which tells you something about his opinion of me.

"So the copper, obviously used to handling this sort of thing, said, 'If you were nowhere near that slogan when it was being painted, how come you've got whitewash in your hair?' He was a smart copper, dealing with unsmart miscreants. They took our names and addresses. I don't know whether we were put on bail then or just told to piss off home. Anyway, they found the whitewash near where the slogan was, and near where we had been seen nonchalantly whistling our way up the road, and we got called before a magistrate in Bradford. I knew the fellow who was presiding, because he was on the executive committee of the Trades Council, and I was a delegate. But he didn't respond to my friendly overtures. In fact, one of the constables said, in a grim voice, 'Take your hands out of your pockets'.

"So they read out the charge and we pleaded guilty because there was no sense in not. My friend had the forethought to tell them what the message was going to be, had we completed it. I made a statement saying we firmly believed in the message. We got charged a fine of £5, a huge sum of money (it was probably more than a week's wage for your average worker), and fourpence cost, because we pleaded guilty and didn't call witnesses, so they couldn't rack up the costs too high. We were given a little time to pay.

"A few days later I was walking somewhere in Shipley and I came across my old scoutmaster and he said, 'I heard you were caught...' He said, 'What was the fine?' I told him it was £5, a huge sum of money, and fourpence cost. He said, 'OK, we'll do a collection.'"[17]

Bill's old scoutmaster was chairman of the Shipley branch of the CPGB.

---

17 Bill Leader, interview with the author, April 1, 2014.

"This is me inciting a crowd..."

612 Craftsman Leader
(collection: Bill Leader)

I would describe Bill's characteristic tone as light, gentlemanly and irreverent. Gloria recalls him holding forth on the soapbox in the carpark on Broadway, Bradford's answer to 'Speakers' Corner'. No-one now remembers his theme, but Lou took exception to his facetious tone, and gave him a dressing-down. Bill puts drollery before dogma and has scant respect for sacred cows. The *Daily Worker* is probably the holiest of bovines. Lou, you may remember, distributed underground copies when the paper was banned in 1941. Jo recalls "going round the houses" of Shipley on Saturday mornings delivering *Daily Worker* with her aunt. It was peacetime and it was legal again. Jo points out that the paper was popular among the racing fraternity because of the reliability of its tips.[18] Gloria and her best friend Heather sold the *Daily Worker* on the streets of Bradford. She has this to say: "It's the only daily Socialist paper produced in England – if you're a serious Socialist, you read the *Morning Star*."[19] Note the change of name: the newspaper became the *Morning Star* in 1966. Gloria's sentiment was forcibly articulated by John Hasted in a song from an old *Sing*: "You must read the *Daily Worker* if you want to beat the boss."[20]

Bill's tone is sceptical, tinged with the oblique.

"It was a very strange change of name, because the Morning Star, I think, is the one that disappears when dawn arrives. Maybe they weren't expecting to be around come the revolution..."[21]

It was the kind of remark that used to drive Lou to despair.

With the end of apprenticeship came the end of deferment of conscription. The scenario was familiar: lads not old enough to vote (the voting age was lowered from 21 to 18 in 1970) received perfunctory training, starvation wages and the thankless task of propping up an empire in collapse. Six weeks of "square-bashing in Dorset" was deemed sufficient to face armed revolt in Kenya ('The Feast of the Mau

18 Jo Pye, interview with the author, December 11, 2013.

19 Gloria Dallas, interview with the author, July 20, 2016.

20 John Hasted, 'Ballad of the Daily Worker', *Sing* Vol. 1, no. 6, p. 120. Sample verse: Now we print the truth on wages, we print the truth on peace / And the truth about the tactics of our wonderful police / With Johnny Campbell at the helm the Worker leads the way / We're out to smash the Tories and we'll smash them every day. CHORUS: So, shout it on the rooftops that all can understand / The workers' daily paper is the finest in the land!

21 Bill Leader, interview with the author, November 6, 2015.

Mau', as Screamin' Jay Hawkins tastelessly put it), civil war in Cyprus, 'Emergency' in Malaya (the term 'war' couldn't be used for insurance purposes) and the mandate in Palestine. Hong Kong, by comparison, was only medium hot.

The territory had only recently been restored to British rule. Renewed conflict between nationalists and communists in mainland China had resulted in sufficient human displacement to create an economic miracle, as ever, powered by cheap labour. Amateur footage shot in 1952 by Michael Rogge, a Dutch banker stationed in Hong Kong, shows the colony still in the underdevelopment phase: peasants quarry in a stone valley – a pretty girl chips away, catches the eye of the cameraman and shyly smiles; scaffolding in Icehouse Street is as rickety as the endgame of Jack Straw's Castle; naked children squat in the streets.[22]

This was the destination of 612 Craftsman Leader of the Royal Electrical and Mechanical Engineers.

"I'd got a trade as an instrument maker which was one of the nicer things to be in the army. I ended up being what they call a Light Aid Detachment attached to a Royal Artillery regiment looking after whatever they had to look after, which was mainly binoculars. Anything that was a bit delicate and wasn't a ten-ton lorry was considered an instrument, more or less."[23]

A "sincere lad"[24] of a deep red hue, Craftsman Leader was predisposed to side with freedom fighters in any clash with colonial force. It was a foible that caused his superiors to take a special interest in him.

"The corporal who was company clerk used to come back to our small billet, and say, 'Intelligence has been down again, checking that you're still here.' I was the only politically committed person, but others had a rather more liberal turn of thought. A bunch of us started looking at what there was available to be interested in in Hong Kong, other than finding a prostitute, which is what we were encouraged to do. Prostitutes lined up on the way to the barrack-gates, so you had to

22 *Hong Kong as it was in 1952* by Michael Rogge, sourced at https://www.youtube.com/watch?v=hcAmRWoH3fQ.

23 Bill Leader, interview with the author, March 15, 2012, and for most of the following Hong Kong quotes, interspersed with some material from another interview, April 17, 2014.

24 Bill Leader, interview with the author, July 31, 2019.

pass a whole line of them when you went out, or when you came home. Because it's better to go screwing some whore than it is to get to know some local girl who you must later be prepared to shoot. We were there to make sure that the local population didn't get too uppity, and didn't want their independence, or anything deplorable like that.

"Each regiment had a riot drill: you formed a square of soldiers, each file looking outwards, to the left, to the right, to the front and to the rear, and the point of the drill was to proceed up the street in a square, with rifles pointing outwards. Everybody had a rifle, except for the fellow who carried a bugle. Oh, and one other fellow. And on the command from the officer, the bugle player would play his bugle (that's interesting: I don't know what he played on his bugle), he played his bugle, and the other person who didn't have a rifle had a placard, which he held up and which said in English and Chinese, *DISPERSE OR WE FIRE!* We were there to make sure they dispersed. And we fired if they didn't.

"It wasn't a defensible place. Just occasionally they got a bit uppity, because the other side of the border was communist, and the Korean War was on. Every now and again we formed up in the mornings and the sergeant would say, 'Anybody who wants to volunteer for Korea, take two paces forward.' One time everyone took two steps back leaving me standing alone. They all laughed.

"So yes, we took an interest, this group, about four of us, going around looking at things that we'd not seen before. You could walk into a restaurant and get all sorts of interesting food. Chinese Opera was performed on street corners. The bookshops were fantastic. It was a propaganda centre. American stuff was all around, but so was a lot of Russian stuff, and Communist Chinese stuff. You could buy books, all in English, that you couldn't get anywhere else. And the pornography was absolutely hair-raising, until it all went off the streets because the Duchess of somewhere-or-other was due to visit, and they decided that it shouldn't be available in case she should walk down the street and ask to see some pornography. Not that she was likely to.

"The main eye-opener was, talking to people in my little group on the troop ship going out, they all had an avuncular attitude towards the people that were going to be around us when we got there. But within a week or so of us landing, they were happy to kick a chink. They had complete contempt for the locals. They showed it at every opportunity.

Again, it was an attitude not discouraged by the authorities, because if you get too friendly with these people, you might not want to shoot them. So that was depressing. But it was enlightening too."

The length of conscription was two years, but "they encouraged conscripts to sign up for three years: the chance of a lance-corporal's stripe being the lure".[25] Bill contented himself with the minimum. He re-entered Civvy Street in 1953, took up his old job and promptly quit.

"I didn't find any engaging employment, and not a lot of useful employment. Because I was involved in other activities. I started selling records for this bunch of nuts in London called the Workers' Music Association…"[26]

Alex Eaton, a charismatic teacher of French at Cartlon Grammar School in Shipley, was his mentor. Born in 1923, Alex inherited his politics from his maternal grandfather, Grandpa Green, active in the early Labour movement and a board-member of the Co-op British Shop in Bradford. Alex, aged eighteen months, contracted polio, which affected his left side, particularly his leg, but fortunately missed his arm and lungs. The surgeons transplanted a nerve from his hip to his foot, so he was able to raise it without it trailing. He was left a severe limp, which he controlled with rapid movements and a gait that gave his step a forward rolling motion. This might have made Alex the target of playground mockery at school, but no, the arrangement was two-way: inspiration in exchange for respect.

He willed himself to climb hills and swim and cycle – Alex and his father would cycle to Blackpool and back – and the gruelling Aldermaston marches were undertaken with something like exultation. Among many other attractive personal qualities, Alex possessed a lovely speaking voice.

"He had a local accent, but he sounded all his consonants. He was a very precise speaker, but in a local accent, and not so much dialect – he didn't use strange words – but he certainly had a very firm, proud local accent, which many a schoolteacher would have tried to mask over the years."[27]

25 Bill Leader, annotation, April 6, 2020.

26 Bill Leader, interview with the author, March 15, 2012.

27 The quotes from Bill on Alex Eaton are culled from two interviews with the author – April 4, 2012 and October 5, 2012.

The platoon, 1952, with Bill standing, middle; Kenneth Williams impersonation; Craftsman Leader and typewriter; three soldiers with Curly Woodacre, middle; "I think he was probably the first Lancastrian I ever talked to. He spoke really funny. Being from Yorkshire I could spot anybody who spoke funny. He was quite a character and I did make attempts to keep in touch with him" (collection: Bill Leader)

Bill singled out four people for special thanks when he accepted a Lifetime Achievement award at the 2012 BBC Folk Awards.[28] Alex Eaton was foremost among them.

"He was the sort of person that encourages you and enthuses you, and believes in you, and gets you to believe you can do it," he elaborates.

The two friends established a branch of the WMA in Bradford in 1954. Camilla Betbeder from the Bishops Bridge Road office in London shipped them discs, mostly 78s, from WMA's record label, Topic. Shellac, a fragile medium, was liable to shatter on touch, and many didn't survive the rigours of transit. Print-based material was more robust. And Alex founded a singing ensemble based on the WMA Singers called… the WMA Singers. Bill and Gloria were members. Bill conducted on occasion.

"Alex's job was to teach us how to sing. He was the only one who could read music. Then I stood in front and waved my arms in a vaguely rhythmic fashion. Yes, without any knowledge of music or anything like that, I conducted the choir."[29]

Alex made the connection between the choir and practical revolution:

> The key is action. We learn as we come to grips with a problem. Activity that involves people and raises them from passivity, no matter how modest at first, will have revolutionary consequences. People can reject efforts to reduce them to obedient ciphers. So, I thought, my projected choir might, if it grew large, be encouraged to expand into other creative activities, maybe even support certain broad forms of political action. What a dreamer![30]

---

28 Reg Hall, Helen Leader and John Ellis were the others.

29 Bill Leader, interview with the author, February 27, 2015. And, further: "The more I'm driven to recall my career as a conductor, the more ludicrous the episode seems. I had no knowledge of music theory. I knew nothing about conducting, except that I had once read a chapter in a book about beating time. On the other hand I was no great shakes as a singer either. I suspect that Alex – who could make some sort of sense about all those tadpoles stuck on those telephone wires, and played the guitar to boot (a rarer skill in those days than now) – thought the best position for me was to have me with my back to the audience, waving my arms about and, some would, say, drowning." Bill Leader, annotation, April 6, 2020.

30 Alex Eaton, *Dot and Carry One*, pp. 1,777-1,778. *Dot and Carry One* is Alex's autobiography, written in the last years of his life for his grandchildren, and can be accessed as an e-book.

"Alex was a great believer: it wasn't just all politics, it was about enjoying yourself, and singing songs that had a message," says Louise.[31]

Individual trade unions had converted manor houses into convalescent homes in the past. The Co-op had turned a small country house at Roden, near Shrewsbury, into a holiday home for workers as far back as 1901.[32] The purchase of Wortley Hall by the AEU, however, took role reversal to a new level. The former seat of the Wharncliffe family opened under new management in May 1951 as an educational centre for workers. The event inspired Bill's one and only song. The family was offended. It seemed to be rubbing it in.

A distinguished and eccentric Wharncliffe, Lady Montagu, wife to the ambassador to Turkey, insisted on her infant son being injected with live bacillus and became an early champion of vaccination. Another notable ancestor, the first Earl of Wharncliffe, was chairman of the Great Central Railway. Debasement came with his descendant, the raffish rock 'n rolling fourth Earl of Wharncliffe. Not that the family's ruin can all be laid at his door. The requisitioning of the Hall during the war, and its rough handling by the army didn't help. The Labour government delivered the *coup de grâce* – death duties of up to 75 per cent on large estates. The ancestral pile was placed on the market. Bill marched under their banner of the buyers – the Amalgamated Engineering Union – on May Day, 1951. Vin Williams, an AEU organiser, was the first to spot Wortley's potential as a retreat for workers. Vin was such a fervent believer that he called his son Lenin.

Such is the context of 'The Ballad of Wortley Hall', a product of the WMA Summer School in 1954, and inspired by the venue where the event was held. The lyrics were the joint work of Bill and John Hasted, and John set them to the tune of 'The Limerick Rake'. I opt for a straight recital in front of an audience of Louise (she has heard it before) and Felix Eaton. 'The Ballad of Wortley Hall' is a demanding tour de force, requiring a broad Yorkshire accent to do it justice. Felix, it seems to me, is quick to pounce on errors of pronunciation.

---

31 Louise Eaton, interview with the author, February 25, 2015.

32 Clare V.J. Griffiths, *Labour and the Countryside: The Politics of Rural Britain 1918-1939* (Oxford University Press, 2007), p. 298.

'Ave you ever been to Sheffield Town?
Thar's burnin' chimneys fa miles aruun'.
And t' smoke curls up and t' smoke comes doon,
And it settles its soot all o'er the toon.

T'were back in nineteen fifty one
A hard day's work were o'er and done,
We were sitting aroun' with nowt to do ["emphasise the hard 't' of 'nowt'"]
When Vin Williams walks into t' AEU...

"Who is Vin Williams?"
"I don't know the name."

Ten miles from 'ere there stands a 'ouse,
No one there but a rat and a mouse,
T'lords and t' ladies have all gorn awa',
T'grooms and t'footmen no longer sta'.
'Ere we sit in t'smoke and t'grime ...

"You don't say the 't'. It's 'in ... smoke'. It's almost a silent 't'. There's a slight hesitation that tells you there's a missing 'the'." Felix makes a guttural noise. "Whereas if you said 'in smoke and in grime', it delivers a different feeling, a different message."

"Go on, you're doing very well," encourages Louise.

To take from... t' rich it is no crime.

We'll build a 'ome for owr workin' men,
If tha wants owt for nowt tha mun do it theesen.

"You must do it yourself. '*Tha mun do it theesen*'."

I've been to t'agent and asked about rents
And told 'im straight I'd nobbut two pence...

"Nobbut two pence," Felix repeats, in mellifluous tones.
"I like this," laughs Louise.

His lordship said tak' it, it's nowt but antique
For a bottle of whisky I can give you now,
But the pound a week I'll 'ave to owe,
For a pint of John Smith's Magnet Ales tha mun call...

"John Smith's was a Yorkshire brewery, of course. A famous Yorkshire bitter," Felix informs.

Now I'm the proud owner of yon Wortley 'all
I've seen t'union leaders, MPs an' all,
But they couldn't care less about Wortley 'all.
So come all yer painters and plumbers and spreads,
We'll put mattin' on t'floor and quilts on t'beds.
Well ceilings were down and t'windows were brokken
Tapestries faded and panelling rotten,
Through t'oles in t'roof the sky showed so blue
But we soon bunged 'em up when t'lads all set to.

"Brilliant," is Louise's verdict.[33]

In 1958, Bill's fifth Summer School, he reprised 'The Ballad of Wortley Hall' at the final concert, with Alex Eaton in place of John Hasted… This is where Louise heard it.

"I mean you don't waste a thing like that," says Bill.[34]

The performance in 1954 had dramatic consequences.

"The Earl of Wharncliffe turned up," Bill relates. "He'd been around the whole week. He was enjoying all these various people doing their musical best, and then we got up and sang 'The Ballad of Wortley Hall', and the rock 'n rolling Earl of Wharncliffe got up in disgust and walked out. As they say, the pig got up and slowly walked away. He was known as the rock 'n rolling Earl of Wharncliffe in the *Daily Mirror*. He was a bit of a rogue. I think it was he who had pissed the family fortune away and why they had to sell Wortley Hall. And he used to drive through the village at a very high speed in very sporty cars and some constable had the temerity to stop him. I think he ended up in court or something."

In 1979 he killed a woman in a head-on collision whilst drunk ("that

---

33 Louise and Felix Eaton, interview with the author, February 25, 2015. 'The Ballad of Wortley Hall' was printed in *Sing* vol. 3 no.3 (Aug.-Sept. 1956), p. 38, and reprinted in *The Bulletin* (the WMA newsletter), March 1958.

34 Bill Leader, interview with the author, March 16, 2020. The following Wharncliffe-related quotes come from separate Bill Leader interviews – February 27, 2015, and March 16, 2020. Sources for the Wortley Hall story are http://www.thestar.co.uk/news/worts-and-all-story-of-labour-s-home-1-3515221 and http://people.com/archive/richard-wortley-a-yankee-from-maine-joins-britains-ruling-class-as-the-new-earl-of-wharncliffe-vol-27-no-26/.

could well be, it's all the one to him") and recovered from his injuries sufficiently to serve six months in prison.

Bill didn't only upset the local lord. It seems Vin Williams had left Wortley Hall under a cloud the previous year, and the subject struck a bum note with the management team.

"So it didn't go down too well."

The story has a postscript. Upon the death of the fourth Earl of Wharncliffe in 1987, the title passed to Rick Wortley, a 34-year-old construction worker from Maine, USA. He received the title and precious little else. Today Wortley Hall maintains its link with the unions, although its core activity has shifted from education to weddings and civil partnership ceremonies.

Bill's work experiences up to 1954 had been monotonous and dispiriting; the common lot in the fifties. The conventional world of work held little appeal. Whereas no job was half as thrilling as selling records for "this bunch of nuts in London called the Workers' Music Association." When, in September 1956, Alex organised the first session of the Topic Club in Bradford (currently the oldest active folk club in the UK), Bill was not present. He had been in London since the previous April. This, following a casual conversation at a screening of *Battleship Potemkin* at the New Era Film Society. New Era, a national network of film clubs, specialised in films from Europe denied general release in the UK: that is to say, all of them. Eisenstein, Pudovkin, with Jacques Tati for light relief, were staples. It might be said that the advent of Eisenstein marked the end of Bill's blind alley days. Cinema was about to change Bill's life for the second time.

The boy… is quiet and sympathetic, mutely observing, always resisting. He must get to know everything.

RICHARD WINNINGTON ON *THE CHILDHOOD OF MAXIM GORKY, FILM CRITICISM AND CARICATURES, 1943-53*

VII

In the Dark

Bill acquired the cinema habit in Dagenham. His first film was "a strange Gracie Fields thing before Ealing found their form". Bill is sure it wasn't *Sing As We Go.* Consulting a filmography of the popular singer and actor, *Queen of Hearts* from 1936 fits.

"For a period, I had a regular dose of film when some enthusiast projected silent films in the scout hut. I remember it as being on a weekday, in the early evening, and a long walk away, for a lad of seven. I saw *Rin Tin Tin* and all those. Ah! It was great."[1]

1 Bill Leader, interview with the author, October 10, 2014.

His first film after the move north was *Pinocchio*, which he saw on its release in 1940. Somehow, he missed *Fantasia*, which preceded *Pinocchio*. The war was over by the time he caught up with it.

"I really struggled with *Rite of Spring*. And Stokowski shaking hands with Mickey Mouse… I don't know. The time I'd seen him before he was shaking hands with Deanna Durbin, so that was a come down."[2]

All the Leaders, with the partial exception of Bill Snr, were keen film buffs. On Mondays, Jo would come home from school and find a sixpence on the table. This was a signal for her to join her mother and aunt in the sixpenny stalls at the nearby Saltaire Picture House.

(Ah, you ask, but how did Jo and Maggie and Patrick come to be in Shipley, when we left them in Canning Town? Briefly put, Maggie was working for a tailor in Whitechapel at the outbreak of war, and Patrick was in the Merchant Navy. He insisted that Maggie and her sister move to the safety of Shipley, where they had Bill Snr and Lou. The sister soon returned. "She couldn't stand the quiet," says Jo. Maggie lasted longer and Jo was born in Shipley on August 23, 1942. She wasn't there for long and was baptised in Canning Town six or eight weeks later. Then, in 1949, Patrick was employed to install the heating system at Bingley Training College, West Yorkshire. It was a long-term contract, and Shipley was attractive beside the open bomb site of Canning Town, so a longer period of residence followed. Jo went to school there, and very often to the pictures. Then Maggie missed her family and she and Patrick and Jo moved to Hoddesdon in Hertfordshire, which is how Jo happened to be at the Hoddesdon Folk Club when President Kennedy was assassinated. Bill heard the news at his own folk club, The Black Horse Broadside, but we're anticipating.)

The beauty of cinema is that strangers can sit in the dark and dream in unison. Some, like Maggie, ignore the collective and dream independently.

"Aunt Lou would go home to get the tea ready, and mum would sleep through the programme for a second time," remembers Jo.[3]

2 Bill Leader, interview with the author, April 1, 2014.

3 Jo Pye, interview with the author, March 20, 2018.

The ritual was repeated on Thursdays, when a new film opened.[4] Two programmes per week couldn't slake Bill's craving, however, and he wandered far afield for fresh celluloid adventures. Bradford was handsomely provided with cinemas, so Bill would spend a free afternoon at the New Victoria, the biggest cinema in the north, and return home in the evening to Saltaire Picture House. I lazily called Saltaire Picture House a 'fleapit' in an early draft and got a flea in my ear from Jo. For the record, the Gaumont, known locally as Saltaire Picture House, was a very smart cinema. It just couldn't compete with the huge, fabulously decorated picture palaces in Bradford, with their mirrored corridors and plush carpets.

"New Victoria had the organ come out of the floor, and it was a genuine, accept-no-imitations Wurlitzer. Norman Briggs played it. He was a broadcasting organist of the era. He was a nimble lad. He used to come out of the floor, playing something fairly vigorous, and he'd swing around on his seat: lift his feet up and swing around. I think he had a microphone. I think he talked to us. And then he'd play whatever it was. It wasn't a very long interlude. It usually included an imitation of a steam train, and the other things that Wurlitzers could do."[5]

Gloria tells how Bill would bunk off school to go to the pictures, and, in a notebook consecrated for the purpose, would jot down the particulars of director and actors, not excluding composer and conductor, with critical comments beside each entry.[6] The film that changed his life was never going to be as obvious as, say, *Citizen Kane*. No 'Rosebud'. No snow globe. I didn't know about its existence, and I dare say the memory was submerged in Bill's consciousness, until an innocent remark brought it to the surface.

"Oh, I saw *Love on the Dole* last night."

"Apparently it was heavily censored," returns Bill.[7]

The 1941 screen adaption of Walter Greenwood's novel boldly depicts a scene of police violence against workers. This is less incendiary than it

4 "I remember an advert appearing in at least one of Shipley's local papers (we had two at the time); summarising the two features that would be shown in the course of that week as 'Two Thousand Women followed by The Hairy Ape'. Eugene O'Neill didn't think of that twist." Bill Leader, annotation, July 1, 2020.

5 Bill Leader, interview with the author, October 10, 2014.

6 Gloria Dallas, interview with the author, February 27, 2013.

7 Bill Leader, interview with the author, February 13, 2015.

might have been, because the police have been provoked to retaliation by rampaging hothead John Slater. In other words, the rabble started it and are getting their just desserts. This is a falsification of the scene in the book and the incident that inspired it, the Battle of Bexley Square in Salford on October 1, 1931.

Whilst taking a stroll in Salford the other day I chanced upon Salford Magistrates' Court and noticed a plaque on the wall. Headed 'Battle of Bexley Square', it read: "This plaque is laid to commemorate the 'Battle of Bexley Square', which took place on 1 October 1931. A poignant moment in history where Trade Unionists rallied to protest against injustice to the working-class people of Salford."

Jimmie Miller (Ewan MacColl) was active in the National Unemployed Workers' Movement at the time and present on the day, as was Walter Greenwood, the author of *Love on the Dole*, and Paul Graney, the folksong collector, and Edmund Frow, the founder of the Working Class Movement Library in Salford. It happened that a protest organised by the Salford branch of NUWM against the hated Means Test was attacked by the police. It was not Peterloo because no-one was killed, but it was bad enough, with demonstrators 'kettled' by the police and attacked and beaten. Paul Graney was among those herded into a cul de sac by the old hospital and was only saved from a savage beating when the nurses opened the windows of the hospital and started pulling men through. Frow received a broken nose and a five-month prison sentence for his part in the day's events, and it provides an exciting scene in *Love on the Dole*, except, as I say, the film is a misrepresentation. The lapse excludes *Love on the Dole* from contention as Bill's life-changing film, but it provides the breakthrough.

"Can you look up *Love on the Dole*? Because the fellow who appeared as the young working-class laddie, with his peak cap and everything, appeared about that time in lots of films, and I don't know his name. I wonder what the hell happened to him?"

His name is Geoffrey Hibbert (this, after targeted clicks). He is Harry in *Love on the Dole*, the brother of Sally (Deborah Kerr, an exotic bloom in the slums of Salford). Hibbert is best known for *In Which We Serve,* but, as we know, Bill's verdict on *In Which We Serve* was harsh. The actor, however, made another film in 1941 with *Love on the Dole* director John Baxter. That film is *The Common Touch.*

"*The Common Touch* was a strange thing. The credits were shown over Tchaikovsky's *Piano Concerto*, and Mark Hambourg, a concert pianist, actually appears playing it in the street on a piano with wheels."[8]

Thus identified, I'm able to catch up with *The Common Touch* on the estimable Dubjax channel on YouTube (most, if not all, the films discussed below have appeared on Dubjax).

Strangely, there is only one reference to the war, when a resident of Charlie's, a refuge for the homeless, or, in colloquial terms, a dosshouse, rants something in German. "He's had a sticky time," explains Tich (Edward Rigby), the odd-job man and knocker-upper. "'E didn't quite see eye to eye with Master Hitler." Yet the film is preoccupied with one vital question: *what sort of society do we want when this bloody war is over?* Lest this seem overly didactic, the pill is sweetened with lots of delirious musical scenes.

What really marks *The Common Touch* as a wartime film is its sense of the fragility of things – all we cherish might soon be lost. The opening scene, for starters, crams in cricket, the playing fields of Eton and Geoffrey Hibbert.We first see Peter Henderson (Hibbert), the school's team captain, driving the ball over the boundary on Eton sports day. The match is his swansong: Peter, an orphan, is due to leave the school on the morrow to take over his late father's City firm. The cricket steward, played by Bernard Miles in thick spectacles, gives an inspiring speech: "This young gentleman's played a good game this afternoon, but tomorrow he's got to start out on life, and life is the hardest game of all. It isn't always easy to stand up to the bowling, and it isn't always easy to keep a straight bat…"

Charlie's too is endangered. The threat comes not from Germans (the film predates the Blitz), but from Peter's City firm, which plans to develop property on the site. A resident of Charlie's, Ben (Bransby Williams), a pure-hearted tramp, hoards yet more endangered objects: cheap geegaw toys, which he picks up on the street. "Fashions all change, you know, and someday people will turn again to simple, beautiful things," he says, extracting every last drop of pathos from the line. Cartwright (Raymond Lovell), a shady director, is relying on the callowness of the new boss to push the scheme through. Yet Peter's curiosity is piqued by Charlie's, and he and a friend masquerade as homeless youths to find out more. A second education follows as Peter

8 Bill Leader, interview with the author, March 15, 2012.

comes to love Charlie's and its residents, a revolving door of drifters, misfits, resting musicians, and shabby lawyers.

A film that started as a ripping yarn about privileged toffs has turned into a drama with a political imperative. *The Common Touch* will accept nothing less than a society where class division is no more. This is succinctly expressed in the last scene when Charlie's inhabitants, safe from eviction, discuss the future.

"All this talk about better things, homes and all that, do you suppose they really mean it?" asks Tich.

"You know, I think they really mean it this time, Tich," replies Ben.

"Blimey! It would be like heaven on earth," says Tich, flabbergasted.

The camera commences a close-up on Ben until his face is so big that the tear trickling down his cheek is a moderately sized rivulet. He turns from Tich to address the viewers in the dark.

"*And why not?*"

The music swells and a title card reads, 'The End'. Ah, but for one young cinema-goer in Shipley this was not the end but the beginning.

There's a lot of music in *The Common Touch* because music is the practical means by which the social transformation is to be achieved. At the risk of over-simplifying, the toffs go for bossa nova with lots of razzamatazz, whilst down-and-outs prefer syncopated, swinging jazz. High and low come together when Mark Hambourg performs the Tchaikovsky *Piano Concerto* on the street on a wheeled upright piano. We segue into a flashback as the orchestra enters, and the names of international cities appear as whizzing title cards, evoking the pianist's glory days. The music subsides, and we return to the hard-up pianist, down but not out. His homeless buddies are awed and enraptured. "The boys like a bit of music," explains Tich to Peter in his bluff, cheery way.

No film more neatly sums up the aims of the Workers' Music Association than *The Common Touch*. It prepared Bill for his life's vocation. Certainly, Tchaikovsky's *Piano Concerto* stirred him to his heart.

"The music really got to me. I must have said so, because my mum bought me the Toscanini/Horowitz version for Christmas."[9]

Was it the music or the message, something about the mutuality of music and social awareness?

Roger Manvell's *Film*, published by Pelican, was another revelation.

9 Bill Leader by email, January 3, 2017.

"It was one of the very early books that dealt with the study of film as a cultural medium, which included history, but also analysis. I read it in 1945, or '46. No, probably earlier. [*Film* was published in 1944.] The other thing that Penguin did after the war was the *Penguin Film Review*. They had these occasional magazines, or booklets. They had a film one and a music one. And I've never come across them in second-hand shops. Not that I've gone searching for them."

John Slater, who starts the riot in *Love On the Dole*, put the 'lumpen' into 'proletariat'.

"God had given him features that didn't really fall eloquently on the face. He looked like he didn't have anyone of any significance in his background. He was in lots of documentaries posing as an ordinary person. He looks as if he walked in off the street to a film studio. He was the nearest thing to real life that British cinema produced until well after the war."

"Look at all the films he's been in," I say, resorting to Google again. "*Went the Day Well?*"

"Yeah."

"*A Canterbury Tale*."

"Yeeaahh."

"*It Always Rains on Sunday*."

"Yeeaahh!!"

"*Passport to Pimlico*."

"It's steady work that. Look at that… 'partial filmography'. Because with three million credits, it has to be partial."

John Slater was among the residents of Charlie's in *The Common Touch*, and the perjurer who sent John Mills to prison in *The Long Memory*.

"He was a man of the people," avers Bill, "who can appear to be natural in front of a camera, which is the trick. He did it. He looks like an ordinary person and could act like an ordinary person. Unlike that other idiot, Niall MacGinnis, who also looked as if he was the real version except he couldn't manage to speak his parts."[10]

The vehemence of the statement surprises. It's possibly unfair from what I've seen of Niall MacGinnis. He was convincingly horrified as Karswell, the evil magician in *Night of the Demon*, reduced to scrabbling on railway tracks for a windblown bit of parchment before fate overtakes him in the form of a giant demon who gobbles

10 Bill Leader, interview with the author, December 7, 2017.

him down. Most criticism of the film centres on the visualisation of the demon, but I thought it highly effective. I'll concede that his characteristic trilling noise, suggestive, taken with the railway setting, of stagehands pushing an outsize model of foam rubber and fur on squeaky tracks, was unfortunate. But Niall MacGinnis comes off better than Dana Andrews, a bit creaky himself. No, the acting honours belong to Reginald Beckwith as Mr Meek the medium, and Brian Wilde, impressive in a cameo as a gibbering madman. Brian Wilde is the loved actor from *Porridge* and *Last of the Summer Wine*, of course.

Pictures, movies (what you will) unerringly reflect the tenor of the times. If *The Common Touch* projected a hopeful future from the dark days of '41, *The Passing of the Third Floor Back*, from 1935, reflects the despair of a collapsing civilisation. The film mostly confines itself to the interior of a boarding house (apart from one scene which opens out to a boat ride). Good, represented by an angelic stranger, and evil, embodied by an odious *parvenu*, wrestle for the soul of an innocent skivvy. Conrad Veidt is the mysterious stranger. One of his fellow-lodgers, Miss Kite, strongly (purposely?) resembles Veidt in his celebrated role of the somnambulist Cesare in *The Cabinet of Dr Caligari*. Modern audiences might recall David Bowie in androgynous mode. Miss Kite is played by Beatrix Lehmann.

Some souls are just not worth saving, is the film's message. The boarding-house inmates are scarcely worth the game. Miss Kite is not the worst, but she wears too much make-up. If this seems harsh, it was about this time Virginia songster Blind Alfred Reed warned young women about the threat to their immortal souls caused by an excessive preoccupation with cosmetics ('Why Do You Bob Your Hair, Girls?'). As late as 1956 Susan Pevensie was barred from paradise as a consequence of her fondness for lipstick, and the enchanted horn lay forgotten.[11] Beneath its conventional piety, *The Passing of the Third Floor Back* subversively suggests that evil *can* overwhelm good, and the forces of light aren't infallible. The book was among Lou's good reads.

"I didn't realise that they had filmed *The Passing of the Third Floor Back*. It was one of my mum's favourite Jerome K. Jerome books. I don't think she knew either."[12]

---

11 C.S. Lewis, *The Last Battle*, Bodley Head, 1956. My copy is a splendid Puffin from 1969.

12 Bill Leader by email, February 2, 2017.

Consider the case of West Yorkshire's proud son Eric Portman, born in Ackroydon, Halifax, in 1901. Portman is Eddie in *Daybreak*, who we first meet as a barber. Eddie and Frankie (Anne Todd) can only express their love in terse and evasive banalities. Both have baggage from the past, which includes prostitution in her case and a secret life as a judicial hangman in his. The censor then removed all reference to her back-story, making character motivation all the more unfathomable – *just like in real life*. Enter Olaf, a casual hand on the couple's barge. Olaf is played by Maxwell Reed, a cut-price Marlon Brando. As a cipher of brute male sexuality, Reed always signals trouble (Oliver Reed got the Maxwell Reed parts a decade on; no relation, as far as I'm aware). Olaf's seduction of Frankie sets in train the events that lead to her suicide and Olaf's conviction for murder. One of the advantages of being a secret hangman might be to revenge yourself on the man who seduced your wife and drove her to suicide and do it all above board. Or so you might think. In fact, the plot reveal comes early on, as Eddie breaks down. Thus, revenge is thrown away as casually as suspense and all that is left, in flashback, is the inexorable sleepwalk to tragedy.

The bleakness of *Daybreak* is unprecedented in British cinema. The parallels are with continental film: it is *L'Atalante* set in hell. Or Gravesend, the closest the location scout could find. Gravesend is always a signifier of remoteness in English film. *The Long Memory* finds renegade John Mills bolt-holed in a boat on the Thames at Gravesend. That film is bookended by a character singing a traditional song, which reminds that Reg Hall lived in Gravesend when *The Long Memory* was being filmed.

What does it say about Eric Portman that he was so consistently drawn to tormented and emotionally damaged characters? Think of Thomas Culpepper, the deviant JP in *A Canterbury Tale*, who stalks girls in the blackout and pours glue on their hair. It reflects on Michael Powell's perverse streak that he isn't presented as a monster: indeed, his actions are condoned because they boost attendance figures for a lecture series on old Kent (this seems very spurious now). Then there was Leading Seaman James Hobson, the anti-social submariner in *We Dive at Dawn*, the rigid patriarch in *A Child in the House*, and the sex killer in *Wanted For Murder*.

Best of all, Portman was Lieutenant Hirth in *49th Parallel*, the Powell and Pressburger masterpiece. It's basically the plot about someone trapped in enemy territory and using wit and ingenuity to avoid capture (*Rogue Male* being the prototype). The twist is that our hero, or anti-hero, is a Nazi, and the enemy territory is Canada. He leaves a trail of devastation for reasons of ideology and survival. The scene where he announces the New Order to the Hutterite community and expects them to fall in because of their German heritage, is priceless. Isn't it marvellous how Anton Walbrook can be a cipher for supreme evil (as in *Gaslight*) and supreme good? He is the Archers' favourite good German in *Colonel Blimp* and their favourite bad German in *49th Parallel*. Passing faces include Glynis Johns exuding a saintly aura, Laurence Olivier as a plausible French-Canadian trapper, Raymond Massey as a plausible Canadian (he *was* Canadian) and Leslie Howard, who shows reserves of steel behind the refined surface. And here is Niall MacGinnis, Bill's pet peeve, playing an honourable man honourably (I do think Bill is hard on Niall). Meanwhile, Portman recklessly smoulders and sneers, destroying anyone who dares come near him. The forbidding persona reflects a forbidding man, who, with his secret, guarded private life, had plenty to be tormented about.

"I remember I thought of him as a pretty strange sort of bloke, but then having my naivety challenged when a worldly young lady (she had even worked in London), told me, with casual sophistication, that he was a poofter. The news perplexed me. My concept of homosexuality was, at the time, limited to ill-informed, nudge-nudge juvenile jokes."[13]

There will never be another Eric Portman, now that an actor can be homosexual without risking his career or being sent to prison, and who suffered the hell of the method actor before there was a method.

Bill Owen, the actor Bill first knew as Bill Rowbotham, is Ron, the barbershop assistant in *Daybreak* who takes over the business when Eddie goes to live on a barge. In the same year, in *Once a Jolly Swagman*, he is the burnt-out Speedway rider Lag Gibbon, reluctant to leave the peace of a mental home after one motorbike spill too many. *Swagman* incidentally reveals that mental homes in the early days of NHS were warm, hospitable places and offered haven from a hostile world. Today we would call them retreats and be charged the earth.

If the evidence of native cinema is credible, 1947/48 marked the high watermark of post-war pessimism in Britain. The bitter winter of '47 destroyed faith in the reforming Labour government; the summer was

13 Bill Leader by email, September 17, 2018.

of '48 was cold and drab; the Olympic Games in London were known as the 'Austerity Games' because rationing was still in force, three years after the end of war. Ben's heaven on earth felt like a long hard slog.

Jim Boswell got his break in film posters with 'Camden Girls'. Picture it: an archway, scruffy terraced houses, railed balconies, a pub, a pair of lovers, a spiv, and, dominating the foreground, a pair of huddled girls, one wearing a head scarf and the other sporting a red bouffant. They belong to that desirable and tainted species, the barmaid. Boswell's painting was turned into a poster by the simple expedient of adding a weather-beaten handbill announcing, 'Ealing Studios Present Googie Withers as the ex-barmaid...', and the strap-line, 'The secrets of a street you know'. The film was the 1947 Ealing drama *It Always Rains on Sunday*.

"That's the way Ealing Studios worked," says Sal Shuel, Boswell's daughter. "They had this guy called S. John Woods who commissioned all the posters for years and years and years, and he got everybody: John Piper, James Fitton, Ronald Searle, all sorts of people. Anybody who was really, really good he got to do posters."[14]

If Ealing comedies were breathtakingly exciting, Ealing dramas scored high on wit and observation. Transgression offers the only escape from drab quotidian existence in *It Always Rains on Sunday*, where the stress-levels of the average housewife are roughly equal to those of a prisoner on the run. As in *Ulysses*, the action takes place on a single day (there are judicious flashbacks) – on Sunday, March 23, 1947. The date is on a newspaper someone is reading.

Rose Sandigate (Googie Withers) is hiding her ex, Tommy Swann (John McCallum), an escaped con, in a cramped terraced house, and has contrived to get the family out of the way. Alfie, her youngest, meanders into a music shop where he casually blackmails owner and incorrigible skirt-chaser Morry Hyams (Sidney Tafler) about his affair with big sister Vi (Susan Shaw). A harmonica buys his silence. This is but a minor thread in an intricate network of familial, sexual and criminal ties, which adds to a recognisable portrait of a community riven with desperation. In short, 'a street you know'.

Disappointingly, Jim Boswell didn't visit the set of *It Always Rains on Sunday*, and so passed up meeting Robert Hamer, its dipsomaniac director, Jack Warner and Alfie Bass.

---

14 Sally Shuel, interview with the author, October 17, 2017.

'Camden Girls' by James Boswell. The sketch – from Lilliput magazine – is more expressionist than the poster (courtesy of Sal Shuel)

"He wasn't interested in that," says Sal, matter-of-factly.

Subsequent Ealing commissions included *The Blue Lamp* (1950), *Pool of London* (1951), *The Gentle Gunman* (1952) – a negligible film, but an early scene shows IRA-man Dirk Bogarde planting a bomb on the London Underground in the pre-war IRA bomb campaign that gave young Bill the heebie-jeebies – and a documentary, *The Conquest of Everest* (1953). By the time of *The Rainbow Jacket* (1954), in which banned jockey Bill Owen mentors a youngster to become the next champion, the bottom of the barrel is being soundly scraped. Boswell's commissions petered out, and so did Ealing Studios.

Was there any correlation between the quality of the film and the quality of the poster?

"No, it wasn't anything to do with that. He was just asked to do a poster and he did it. The posters got boring. Things had to change, as things always do. He was given a collection of stills from the film and told to do a poster from them. He didn't like doing that. He did it professionally and efficiently, but not well. The early ones were much, much nicer," says Sal.

Bill is expanding (if that's the right word) on 'stretching', a mechanical method used to restore natural movement to silent films. The speed of early film varied and hovered around the Lumière-prescribed 16 frames per second. There came a time, however, when no projector survived to screen those flickering shadows. Figures moved like nothing in nature when projected on sound era equipment at the industry standard of 24 fps.

"It might be in Google," he says.

'Stretching' and 'film' brings up clingwrap, so Bill has to explain the process himself.

"They would print frame number one, frame number two, and maybe frame number three, and then frame number three again. So you were taking up more time. A picture comes down and is held in the frame very briefly and then taken away, and the next one comes in. What you're seeing is a series of snapshots, but the eye doesn't notice that."

I understand this perfectly, but Bill seems dissatisfied. "That's not very precise," he says.

"*Battleship Potemkin* was the only stretched thing I ever saw. I remember one shot of waves coming in, to Odessa, I presume.

The frames were virtually identical, because one wave coming in is very much like another, but instead of it being a continuous movement, at some point one of them was repeated, and you got this spectroscopic effect, when moving things appear to stand still."[15]

Film started as an innocent entertainment devoted to frantic, knockabout business designed for laughter. Sergei Eisenstein had more serious intent and used shock tactics to inspire revolutionary fervour. Eisenstein orchestrates atrocity with sharp edits; someone has counted 1,346 cuts in the film. Who can forget the Cossack bullet going through the head of the prim little lady, or the pram careening down the Odessa Steps, complete with screaming baby, or the phallic canons spurting in endless orgasm? This was strong meat, as strong as the rancid meat that initially provoked the 1918 uprising. In fact, the film played to half-empty theatres in Moscow in 1925 whilst packing them in in Berlin. The difference was the addition of Edmund Meisel's machine-age music. In some German cities, censors passed the film but banned the music.[16]

And here's another surprise connection: Hyndman Hall on Liverpool Road, the assembly point for demonstrators in Salford on October 1, 1931, doubled as the headquarters of the National Unemployed Workers' Movement *and* Manchester and Salford Workers' Film Society, a precursor of the New Era Film Society of Bradford, and showed the same roster of films – *Battleship Potemkin*, *Metropolis*, *Storm Over Asia* and other exemplars of revolutionary cinema.

Would *stretching* be a good metaphor for the transformation of Bill's life in 1955? The repetition of identical moments give only an imitation of life, but now sound, vibrancy and movement enter to sweep away stasis. The point of departure was *Potemkin*, a monument to monochrome when Bill went to a screening at New Era in 1955, the heyday of technicolour. Here he fell into conversation with the club's organiser, Ron Ogden. Ron said Films of Poland in London were looking for a film librarian. The meeting was to have fateful consequences. Bill uprooted himself from wife,

15 Bill Leader, interview with the author, February 27, 2015.

16 'Original Potemkin beats the censors after 79 years' by Ronald Bergan, *The Guardian*, February 17, 2005.

home, family and friends and lit out for London.[17] 1955 was Bill's 'hot year'.

"That was the year I went down to London and put behind my provincialism, and suddenly found myself thrown into a completely different world. Not just because it was London, but because of the people I found, that I got thrown in with. That's where I started."[18]

The job at Films of Poland, among other things, entailed recording English voiceovers for the occasional Polish documentary short. This brought him into contact with actors whose names he had earlier jotted in his notebooks.

"There was a chinless fellow called Miles Malleson, an old actor. He always played [with quivering voice] 'doddering old fellows, my word.' He was the hangman in *Kind Hearts and Coronets*."

Ah yes, the poetising hangman whose fondness for his own grandiloquence literally saves Louis Mazzini's neck. Since Bill told me about him, I've been collecting cameos from Miles Malleson. He's Old Joe, the shabby fence who distributes Scrooge's worldly goods in *Scrooge*; the tailor puzzled by Sidney Stratton's specifications in *The Man In the White Suit*; a sinister hearse driver in *Dead of Night*; a buffoonish sultan in *The Thief of Baghdad*; a bumbling courtier in *The Private Life of Henry VIII*; a wool-gathering cleric with a tender spot for Miss Prism (Margaret Rutherford) in *The Importance of Being Earnest*; the contented notary in *The Queen of Spades*; legal adviser to the Lords (Stanley Holloway and Kathleen Harrison) in *The Happy Family* (any film which wafts Dandy Nichols to heaven at the Festival of Britain must count as Elstree Magic Realism); the old boy who wrestles with Mr Polly (John Mills) in an unsteady punt in *The History of Mr Polly*. More? In *Knight Without Armour*, we are invited to laugh at the polemic he spouts as the 'Drunken Red Commissar', when actually they were close to his views in real life; an orchestra conductor in *The Magic Box*; an ineffectual Olympic committeeman in *Geordie*; a myopic failed barroom gallant in *Stage Fright*.

"He was good at playing someone who has been overtaken by the world. He wrote a play about the Tolpuddle Martyrs. He anglicised

---

17 "In defence, may I say that Gloria would have come down with me, but she was finishing her Art Diploma? As soon as she finished. she came down." Bill Leader, annotation, July 1, 2020.

18 Bill Leader, interview with the author, March 2, 2012.

a portion of *Les Miserables* called *The Bishop's Candlestick*, which we used to do at school. He did English adaptations of Moliere. He was an all-rounder. A brilliant lad, who made a living[19] out of appearing to be a duffer."

Miles Malleson appeared in hundreds of British films, all of them worth seeing, if only for the two minutes or so he radiates the screen.

"Anyway, he did a voiceover on one of these documentaries, so I worked with him a bit. And the other one I worked with, again these are only just voiceovers, was… Oh quite a famous actress in her day; she had a sister who was quite a famous novelist. It will come to me… Beatrix Lehmann!"[20]

"And then of course there's Orson Welles doing the Findus commercials. Have you heard that one?"

This is Sean Davies, a recording engineer and Bill's contemporary. Sean worked at IBC and ran the studio at Cecil Sharp House before Bill took it over. He collects outtakes of voiceovers and is very entertaining on the subject of Orson Welles and fishfingers.

"Voiceovers for commercials is one of the most unrewarding jobs in the studio. I used to have to do that kind of thing when I was a junior at IBC. And these people from the advertising agencies are such arseholes. So full of themselves. And these little arseholes come around from the agency to produce this recording of Orson Welles, and they try to tell Orson Welles how to read the script. *They weren't born when he made Citizen Kane!* And so he gets more and more fed up with them, until in the end… 'Orson, the best reading…' (irate Orson Welles voice) '*The best reading is the one I'm giving you now. You're such pests!*' And then he comes out with this, and I think this ought to be on the wall of every manager's office: '*Now what is it in the depths of your ignorance that you think you want?*' We've all wanted to say that to somebody at some time or other, especially in the music business."

An innocent enquiry about artists of his acquaintance elicits a stream of such joyful anecdotage that I shall set it down in this place with a minimum of editing, just as he told it to me.

19 Bill said he made a 'fortune' in the original, but corrected himself whilst proofreading the chapter, explaining, "I've fallen into the trap of thinking that all film actors were rich."

20 Bill Leader, interview with the author, March 15, 2012.

"Not long before I left IBC, I did recordings for Caedmon, the American speech label. Caedmon was a Welsh shepherd, I think, one of the legendary types… These two American dragon women, businesswomen, set up this label using some tapes they'd got from South African radio of Dylan Thomas. And then, to their wonderment, American Congress voted that speech recordings were eligible as educational material, which meant that colleges could spend their grants on them. So that was it. These two women, I can't remember their names now [Barbara Holdridge and Marianne Roney], decided they would record everybody in all the Shakespeare plays. It had to be in London, and everybody meant 'Sir' or 'Dame'. IBC got the lion's share of the contract. Nobody wanted to record Shakespeare except me. 'Who can do this?' 'OK, I'll do it.' It was wonderful. The high point of my career really, with Gielgud, Olivier, Peggy Ashcroft, Richardson, all of these people. And I invented a new way of recording because Caedmon wanted separate mono and stereo recordings.

"Now if you take a stereo recording and you add the two channels together, whatever is in the middle comes up three dBs. So if you've got an ideal balance in stereo you'll have a slightly wrong one in mono. We had half-inch Ampex three-track machines at IBC at the time. The BBC way of doing stereo drama in those days was to group people around a stereo microphone, crossed figure of eight, a Blumlein technique. The trouble is, if someone did that, they'd jump six feet in the stereo picture. And you got people jostling around a mic. So I thought I'm going to have a stage front, with five microphones, and the centre mic goes only to the centre track. The outer mics go to the top and bottom tracks. The left-of-centre, right-of-centre are cross-fed between the top and bottom tracks. This means that I can raise or lower the centre mic without affecting the stereo width. Most soliloquies and so on are centre-stage. So I said to the producer, who was Peter Wood, who worked with Tom Stoppard later on, I said, 'It's a stage-front, and they mustn't turn their back on the audience.' And the actors loved it, because it's stage-front. None of this grouping together around a mic. And it worked very well. Gielgud used to say [Gielgud voice], 'Oh I never listen to my recordings.' Rubbish! The first time I recorded him he came up to the control room and said, 'I've never heard myself in stereo before. Wonderful! Wonderful!'

"Everybody's got a Gielgud story. We were recording, the first day, and Peter the producer is sitting next to me at the control desk. He was very much in awe of Sir John. As you would be, you know. Gielgud was a pussycat to work with. No problems. My general experience, both with musicians and actors, is that the better they are, the more reasonable they are. The worst ones are the ones that are really not that good but think they're wonderful. So anyway, Gielgud is centre-mic delivering the speech, and we're listening. [Makes noises…] *rmmbb hmm bmmp brmmbb bmmp brmmmm…* On the speakers. Peter: 'Sean, what's that?' Now there's such a thing as 'bias bubbles' on tapes. You used to get it on sopranos, french horns, but not usually on male speech. So I'm checking the tape now. No, it's not the tapes, it's not 'bias bubbles'. It's on line-in. It's not the mic. Peter: 'Christ! It's his tummy rumbling. We'll have to do it again. I can't tell him. Do you mind if I blame you?' Blame the engineer, always. So Gielgud comes up and listens and Peter says, 'Sir John, absolutely perfect. Wonderful! But we did have, I'm sorry, a technical problem. Could I ask you to do it again?' [Gielgud voice] 'Oh yes, perfectly alright, actually. My tummy was rumbling. I haven't had any breakfast. I can't boil an egg, you know.' The old bugger knew all the time!

"When I first moved here my local pub was up in Waddesdon. Do you know Waddesdon? It's where Waddesdon Manor is, Lord Rothschild's place. The Five Arrows pub represents the five branches of the Rothschild families. That's their crest. Five arrows. And so at the Five Arrows pub one of the bar-staff was a retired actor, dancer, stage manager, and he had a fund of stories, and one of the favourites was that Gielgud was on in a play at the West End, at the National Theatre, I think it was, and the producer was Peter Brook, and he was having one of his silly periods. As Robert in the pub used to say, 'There were others?' Anyway, for this particular play Peter Brook has gone back to RADA, first year. 'I want you to all come on and be a tree, one at a time.' Him sitting in the stalls. So everybody has to come on, even Gielgud, and be a tree. First year RADA, for god's sake. Anyway, one day he says, 'I want you to come on, one by one, and frighten me. I mean really frighten me.' And they come on and go, 'Aawwuuooeeeahhh' [scary voice]. And it comes to Gielgud's turn and he strides to the centre stage. '*We open in ten days!*' And he strides off."[21]

21 Sean Davies, interview with the author, April 19, 2018.

Once Bill inadvertently recorded Derek Jacobi talking about playing Shakespeare. It was the by-product of a session arranged to record *I, Claudius* as an audiobook and Jacobi was chatting away informally afterwards, unaware the tape was still running. This happened much, much later than the period under discussion, and anyway it qualifies as a Lost Leader, as it was swept away when the bailiffs raided Greetland.

Apropos a TV showing of *I Confess*, Bill declares, "I don't think the sun shines out of Hitchcock's arse, I really don't."[22]

I'm inclined to agree. His films give the uncomfortable feeling that you've been delivered into the hands of a borderline pervert. Hitch can be forgiven when he's good, with *North By North West*, *The 39 Steps*, *Psycho* etc. *The Birds* is a good idea by Daphne du Maurier, competently realised. *Shadow of a Doubt* gets the benefit of the doubt because I've been a sucker for Joseph Cotten since *The Third Man*.

An early Hitchcock effort called *Young and Innocent* is more typical. The film has a very elaborate crane shot, where the camera swoops from a panoramic view of a ballroom to home in on the twitching eye of a drummer in black face, thus revealing the killer. It's bravura stuff. The trouble is, the rest of *Young and Innocent* is very stilted. This tends to be the way with Hitchcock films: silly plots redeemed by splendid set-pieces. When he was bad, he was very bad, as with *The Trouble With Harry* and the execrable *Frenzy*. Would it be unfair to compare *Rope* (made in the forties) with *Birdman* (modern), and suggest that Alejandro G. Iñárritu could show Hitchcock a thing or two about making a film in one continuous take? Yes, it would, considering the technical limitations of the medium in 1948.

Bill and Gloria's first shared home in London was in Leytonstone, Alfred Hitchcock's birthplace, but that didn't make it any better.

22 Bill Leader, interview with the author, April 1, 2014.

We cannot always remember our first glimpse of those who later become important to us. Feeling that the happening should have been more significant, we strain back through our memories in vain.

ELIZABETH TAYLOR, *A GAME OF HIDE AND SEEK*

VIII

William and the Cambridge Spies

It might be worth revisiting the crucial conversation between Bill and Ron Ogden, in Bill's own words. The account in his unpublished memoir, *If I Remember Writely – The Unsound Memories of a Soundman*, begins by filling in the socio-cultural background:

> It was 1955. The year that a 15-year-old African American girl had refused to give up her bus seat in Montgomery, Alabama.[1] The year Bill Haley rocked around the clock. The year television in Britain grew a commercial arm. The year Anthony Eden succeeded – if that's the right word – Winston Churchill as premier. The year Albert Einstein died, and Bill Gates was born. The year Ruth Ellis became the last woman to suffer the death penalty in Great Britain – we hope. The year journalist Christopher Mayhew took mescaline on BBC television, but they didn't broadcast

1 Claudette Colvin, the 15-year-old who came before Rosa Parks.

> it. The year *Lolita* was published; Princess Margaret did not marry Peter Townsend; James Dean became a traffic accident statistic; fish fingers first pointed the way to ever faster food, and the Warsaw Pact was signed. The European War had ended ten years previously. Stalin had been dead for two years. It would be another year before Khrushchev blew the gaff on him. I didn't know it at the time, but for me, the sixties had started. I was 25. It was Easter and I had just arrived in London to work for the Polish Embassy. I was going to work in the film library of Films of Poland.
>
> I'd been recommended for the job by Ron Ogden, the man who ran the New Era Film Society in Bradford. He'd got a message which said: "We need someone kosher to work in our film library". *By 'kosher' the message did not mean someone blessed and bloodless; they meant someone who was a card-carrying Communist party member. I carried such a card, as did my parents* [my emphasis].[2]

Dave Arthur's biography of A.L. Lloyd, *Bert*, describes how the CPGB looked after its own with posts to various communist front organisations, typically 'friendship' societies for countries in the Eastern bloc. I am perturbed and raise the matter with Gloria.

"Was Films of Poland a communist sinecure?"[3]

Her response is disconcerting. Bill's ex-wife laughs and laughs and laughs. Indeed, she laughs so hard I'm forced to abandon the line of questioning altogether. Undaunted, I accost Bill at our next pow-wow ('interview' doesn't quite suit).

"Did you get the job at Films of Poland because you were a card-carrying communist?"

Bill sighs patiently.

"Should I explain what Films of Poland was, and how they operated?

"The Polish Cultural Institute was a rather splendid Regency building at the top end of Portland Place, at the Regents Park end. And it was part of the Polish Embassy, which is also in Portland Place, halfway down. It was part of the cultural exchange. If you let us have a cultural institute in London, so that our folks can come in and spy on you, we'll

2 *If I Remember Writely*, ibid.

3 Gloria Dallas, interview with the author, July 20, 2016.

let you have a British Council in Warsaw, so your folks can come in and spy on us, in a gentlemanly sort of way.

"If anyone was involved in spy work, it was the cultural attache. He was mainly trying to get in touch with emigres who might be persuaded to go back. A very distinguished emigre, Andrzej Panufnik, one of their leading classical composers, had made the dash only the year before, and he was in London. It would have been a very attractive feather in their cap if they got him to go back. But they never did. He's still here. He's still writing. He's just got a new symphony out. Andrzej Panufnik. I thought you would have known.[4]

"In the main, Films of Poland was a propaganda device. It was run by a couple of brilliant lads who sold the idea to the Embassy. They were involved in the film society movement, and they knew that people were desperate to get hold of decent films to show in film societies, and, in the main, you had to hire them from commercial companies who had 16mm versions. Shell would let you have copies of their documentaries about drilling for oil fairly cheaply, but if you've seen one you don't need to see many others. But the idea behind Films of Poland was to have the Polish Cultural Institute lend out these 16mm versions to the film societies – the Polish film industry was producing quite a few significant films. And all the film societies had to do was to find the postage to send the films back. Also available, of course, were newsreels and films of specialist interest. They proclaimed the wonder of Poland, and how much better Poland was going to get under the new regime (the new regime being Stalinist).

"So that was the idea of Films of Poland: propaganda with a smile."[5]

Peter Brinson and John Minchinton, the 'brilliant lads', were a case of opposites attracting. Brinson, a former tank commander at the Battle of El Alamein, produced and starred in the first stereoscopic ballet film.

"He'd been the youngest major in the Tank Corps, and a bit of a hero. And he had an accent like Peter Sellers' Grytpype-Thynne from *The Goon Show*. It was wonderful. Certainly, he had the upper crust accent of the better type of person. And rather a haughty demeanour

4 Sir Andrzej Panufnik, Polish composer and conductor, b. September 24, 1914, Warsaw; d. October 27, 1991, Twickenham.

5 Bill Leader, interview with the author, March 15, 2012.

all round. Not intentionally, that was just an unfortunate way his speeches fell."

Yet John Minchinton was the creative force behind Films of Poland, with Peter Brinson as the front.

"He wasn't pulling the strings, but he was doing the real work. Probably having his strings pulled actually. He knew, and he understood film technically, historically and artistically, and he understood it as a medium. He became an expert in subtitling foreign films. There's good and bad in subtitling, and he was good. I sat beside him through several sessions where he was subtitling one of the Polish features. That was really quite interesting. It had nothing to do with the audio side but more to do with the literary side. He worked from a translation of the dialogue, and he had to work out exactly what words appeared where. He was the one who decided where the titles fell and how they fell and what they said. That was quite interesting."

In the biographies gathered in the appendix as 'Possibly Significant People', I give John Minchington's occupation as 'subtitler'.

"'Subtitler' is a bit dismissive," objects Bill. "I think it skews the biog if you put the stress on eastern European film. After all we have no authoritative explanation of how Films of Poland got off the ground. Is there a point in mentioning the Czech honeymoon? I'm trying to think of some way of demonstrating the range of his talents, and describing him simply as a subtitler, works against that. After all, talking about the little we know (in some cases) and inevitably not mentioning the bits we don't know, must skew the biogs."[6]

No really, this gets to the heart. Some worry that Bill's work as a sound engineer is too minor to merit interest. Literary agents, the commissioning editors of big publishing houses and Audrey Winter lean to this point of view. Whereas I interpret my brief as bringing hidden lives to light, according to George Eliot's formulation on the last page of *Middlemarch*…

> …the growing good of the world is partly dependent on unhistoric acts; and that things are not so ill with you and me as they might have been, is half owing to the number who lived faithfully a hidden life…[7]

6 Bill Leader, annotation, July 7 5, 2020.

7 George Eliot, *Middlemarch* (first pub. 1871-2; The Penguin English Library, 1981), p. 896.

Subtitlers and sound engineers are more hidden than most. Their work is undone if they in any way draw attention to themselves. They are self-effacing and non-intrusive out of necessity.

The subtitler faithfully paraphrases dialogue and judges where to make cuts – a literal translation would slow the flow of action. Something is wrong if the written text overpowers the image. Language is the least of it (translators are not so hard to find). Factors beyond the subtitle's control include words that drop off the bottom of the screen and white subtitles that appear against a light background, rendering the words invisible. Tiny captions are a problem on television, where a trade-off must be struck between deafness and blindness. The UK's culture of subtitles is largely thanks to the BBC, yet post-synched dialogue is a creeping danger, and dubbing is the norm in non-English speaking countries, even if voice is as distinctive as physique. I once tried to sit through *Qué bello es vivir* starring James Stewart and didn't get far.

When I was young there was a slot on BBC2 late on Friday nights called *World Cinema*. This was how I came across Bergman's *Smiles of a Summer Night* and Fellini's *La Strada* for the first time. I was at an age when important truths about life could be gleaned from foreign films, whereas the English-speaking variety (Hollywood, I mean) merely pedalled escapism. Some of these truths were bleaker than I was ready for, and tended to go over my head. Luis Bunuel said everything about the human condition in *Nazarin*, and, although it passed my understanding, I picked up certain clues to follow through.

Images linger in the mind. It has taken a lifetime to track down some, and others elude me still. Bombers swoop to attack a convoy of refugees on a bridge and a little girl is lost. *Jeux Interdits*, or *Forbidden Games*, the saddest film ever made. The girl in her Sunday best who pauses to pat the head of an Alsatian dog as she, her schoolmates and guardians file unknowing into a gas chamber. *Pasezerka*. Or – children again – a Swedish brother and sister attempt the simultaneous swallowing of a whole hard-boiled egg and are overcome with laughter. (I haven't managed to track down this one.) Kids in Scandinavia have it best, was my conclusion. All are Minchington jobs. At a rough estimate, he subtitled 2000 feature films during his working life. Significantly, he left Films of Poland in 1964, the year BBC2 was launched.

(I was able to watch *World Cinema* unchallenged because I was frequently farmed out to an aunt and uncle as an underage babysitter and had control of the TV knobs as a fringe benefit, although my uncle might also tip me a fifty-pence piece – quite new in those days – if he was in a pleasant state of inebriation.)

In the period under discussion (1955-64) Peter Brinson and John Minchington were the joint editors of *Films and Filming*, a fact that was to have some bearing on Bill's sideline as a freelance journalist.[8]

What else went on in the corridors of, well, powerlessness?

"The top floor front, in adjacent offices to Films of Poland, was the print part of the Polish Cultural Institute. Two English journalists produced a periodical called *Poland Today*, or some similar gripping title. They would struggle to render some of the phrases that were the propaganda line of the day into English. They would sit there trying to work out how to translate *The Forward Sections of the Working Class Must Attack the Rear Portions of the Lumpen Proletariat* without the ringing homosexual overtones. Tasks of that nature. Another floor dealt with more general cultural things, like organising exhibitions. The ground floor was a sort of social club, or a large meeting-room that could be turned into a cafe. There was a West Indian lad who did the cooking."[9]

Bill now became a familiar figure in the local post office, dispatching endless packages to film societies. The job was ideal training for the mail order side of Leader/Trailer, a task that would dominate Bill's working life a decade or so later. Then, returning to his basement office in Portland Place, he would check the returns, repair any damage done, and try to keep on top of the admin – whilst doing his best to make sure societies got their films on the allotted date. It was low-level stuff as international diplomacy goes, yet required tact and patience.

"Because there weren't unlimited copies of the films sometimes you couldn't meet a deadline and had to negotiate an alternative date. The deadly thing was, if a new film came out, particularly a good film, we would get perhaps six prints of it, 16mm, and I was told to view the entire film six times through, to make sure there were no faults with

8 "By the way that whole Hanson Books stable of *Dance and Dancers*, *Plays and Players*, *Films and Filming* might be worth looking at." Bill Leader, annotation, November 25, 2018.

9 Bill Leader, interview with the author, April 4, 2012.

any of the prints. The more you see some films, of course, the more you get out of them, but not many films come into that category. So that was a bit of a chore." (In 1955 Andrzej Wajda had yet to make his mark, and *Closely Observed Trains* was still around the corner.)

As work on his biography advances, Bill is constantly being disappointed by people who don't remember meeting him. Leon Rosselson's paradoxical response is typical: "I think I met him without actually meeting him."[10] Sweetness and amiability are fine things, but they must deter from making a forceful first impression. Bill, for his part, remembers his first encounters with many of the important people in his life because they happened on the same occasion – the evening of his first day at Films of Poland.

"I arrived at Films of Poland on Monday morning, all fresh and ready for work, and discovered there was going to be a rehearsal of the London Youth Choir in the ground floor meeting-place that evening. Somebody suggested that I might like to turn up, which I did. That's where I met Leon Rosselson (his two sisters were also involved). He'd finished Cambridge, but only just.

"At that time there was an international organisation called The World Federation of Democratic Youth. They ran a Youth Festival every odd year: 1955 was Warsaw; '57 was Moscow; '59 I think was Vienna, and then there was Helsinki [1962], the only one ever held on this side of the Iron Curtain. Anyway, it gave folks a target to aim for. And the London Youth Choir were greatly bound up in these international youth festivals. They represented an aspect of British youth, albeit a very left-wing aspect of British youth.

"On my first Monday evening in town, I went up to the ground floor. Because I was working there, I could get down quite early and enjoy some of the food they served up. I remember it was the first time I'd ever had stuffed cabbage leaves, courtesy of our West Indian chef. 'This is a brilliant idea!' Then people arrived. John Hasted, who I knew from the WMA Summer School, and a character called Eric Winter. He edited a magazine called *Sing*, which was technically a more modest production than *Sing Out*, and it encouraged people to write songs."

Sing Out kept the US folk community supplied with news, articles, songs and inspiration. Incidentally, they shared office space with

10 Leon Rosselson, interview with the author, March 7, 2013. "I was probably [big sigh] hanging around Ewan MacColl," Rosselson continues.

Folkways Records, circa 1950. The Stateside publication was the model for *Sing*, whose avowed aim was to provide "songs of immediate and topical interest"[11] to members of the London Youth Choir. At the time Eric Winter was a librarian at St Pancras Library; later he became folk correspondent for *Melody Maker*. With John Hasted, he was a major force on the budding scene.

"That was quite a good day for me, because I met a lot of people who were influential in their own way and who were influential on me. That was my first day's work in London. Looking back at it, it was fairly monumental."[12]

No sinecure, then. If the work wasn't done, Bill would be sacked. The pay was desultory. And what of that other famous communist front, the Workers' Music Association? The WMA existed to regenerate society through music. Bill wholeheartedly approved of the aim. What's not to like? And, because the work was voluntary, there was no question of grubby self-interest. Bill would get his reward in heaven, if only he believed in heaven.

At one time that Bill attended a meeting of the music group of the CPGB in King Street. He went there to hear Alan Bush.

"It was a big building on the edge of Covent Garden, opposite a major branch of Moss Bros, the dress hire people. Gentlemen's clothes and things. When you wanted to get a taxi there, you asked for Moss Bros, not the headquarters of the Communist Party. Because the taxi driver knew where Moss Bros was."

Bill reconsiders.

"There were some taxi drivers who knew where the headquarters of the CPGB was, because they were members.

"The thing I attended was Alan Bush talking. I don't know what his overall subject was, but he was being critical of Stravinsky, because Stravinsky said that music can't represent anything. Music was music. So Alan Bush was implying you could write Stravinsky off as being useless because he didn't have the Party line. Someone at the meeting, an Oxbridge professor, made the point at some length that it didn't make a difference what Stravinsky said in terms of how he viewed his music, it was about the music that Stravinsky produced. You can't write Stravinsky off on the basis of his theorisation but on his practice,

11 Editorial, *Sing*, vol.1, no.1 (May-June 1954).

12 Bill Leader, interview with the author, April 4, 2012.

which is much more along the ideas of dialectical materialism, and that sort of thing. Alan gracefully gave way."[13]

The term is 'formalism'. I got it at art school. The idea is that music or the art object must enshrine *form* and resist what is merely literal. Such thinking was inimical to believers in music as a force for social good, like Alan Bush. The Oxbridge professor carried the day.

The chair of the meeting, Jack Dunman, was a Co-op activist, a CP functionary and a big hitter in the National Union of Agricultural Workers. A commitment to the redistribution of wealth, however, didn't preclude a love of music and culture. He believed that opera should be sung in English and fought for it to be official CPGB policy with the energy he had previously reserved for ending the abusive system of tied cottages. NUAW meetings took place at his wife's photographic studio. She was Helen Muspratt-Dunman, whose work combined social documentary and formal experiment; she was fond of natural light but didn't mind experimenting with the solarisation technique of her idol, Man Ray. Helen Muspratt-Dunman made striking portraits of the painter Paul Nash and the dancers Hilda and Mary Spencer Watson. Sitters at her Cambridge studio included Guy Burgess, Donald Maclean and Anthony Blunt, the notorious Cambridge spies.[14]

Like MI5, I'm interested in the connection between Bill and the Cambridge spies. The link is tenuous – a one-off discussion about Stravinsky in the King Street headquarters of the Communist Party – but, as Jo Pye reminds, you don't have to do anything; it's guilt by association. The very existence of the CPGB music group confirmed the suspicions of the establishment – what begins as a taste for radical art must perforce end with the overthrow of Parliamentary democracy.

The 1951 Roy Boulting film *High Treason* is based on the premise that anyone who professes to like new music is either a pseud or a terrorist. They might be ordinary people, or indistinguishable from ordinary people, like Mr Ward (Charles Lloyd Pack), a mild, cat-loving clerk in a docklands office, or Jimmy (Kenneth Griffith), an ex-RAF officer with his own small electrics shop. They come together as members of the Elgin Modern Music Society which promotes concerts

13 Bill Leader, interview with the author, March 16, 2020.

14 Sources: https://grahamstevenson.me.uk/2008/09/19/jack-dunman/ and https://grahamstevenson.me.uk/2008/09/19/helen-muspratt-dunman/.

and produces records, presumably in runs of 99, just like the WMA. Ah, but new music is just a smokescreen for the organisation's real aim, which is to undermine the fabric of society by planting bombs. Mr Ward, making a tardy entrance to a piano recital, is shushed by music-lovers and joins the throng at the back of the hall. The music drowns out his whispered words to neighbour Jimmy: "March the 16th". It is the date of the next terror outrage.

Sir Grant Mansfield (Anthony Nicholls), a character clearly modelled on Tom Driberg, is a sleek, cultured MP waiting in the wings to seize power when the government falls. At the climax, the terrorists mount an all-or-nothing assault on Battersea Power Station. Democracy is only saved thanks to state vigilance and the efficiency of Special Branch's card index system.

The new music avidly devoured by the members of the Elgin Modern Music Society is not so bad. Cinema is the only place where atonality is acceptable, usually as an accompaniment to bloody murder, but here screen composer John Addison sends up Schoenberg without anyone getting hurt. His minor key meanderings are more rousing than intended, and the police infiltrator who almost gives himself away because he can't hide his affront at the Second Viennese School (Alan Bush was Britain's leading exponent) is, you feel, the most awful stuffed shirt. New music was the suspect commodity when the WMA specialised in new music (at the time of *High Treason*); oddly it transferred to folk as soon as the WMA started to take an interest in folk.

Sadly, Bill thinks my Elgin = WMA thesis stretches plausibility. In 1951 the WMA had just resumed issuing records after a ten-year hiatus, and had not yet hit their stride. Nobody at the WMA spoke in the plummy tones of fifties thespians. Bill fastens his full attention on the opening credits as I attempt to play him *High Treason* on YouTube. He spots that Royalton Kisch is named as the director of the New Symphony Orchestra.

"Now that's an interesting character. The New Symphony Orchestra didn't really exist, because it was a pickup group. Royalton Kisch, from what I gather, was a bit of a dilettante. Not really a musician, but very rich, or his father was. Decca had him conducting the New Symphony Orchestra for various things, but they made him pay. As far as I ever heard, it was a vanity job."

(Bill's childish notebooks, filled with his musings on cinematic matters, evidently extended to conductors as well as the more obvious director and actors.)

Oh joy! Peter Jones plays the part of a musicologist, demonstrating his superior sensibility with blazing breeziness. He says things like, "beneath the cerebral there's loads of lyrical." If you look closely, you can spot Dora Bryan, Alfie Bass and the inevitable Sam Kydd.

"Who is the director?"

"Roy Boulting."

"Oh yes," says Bill, noncommittally. "Is it any good as a film?"

Entertaining hokum. The films of Roy and John, the interchangeable Boulting brothers, range from the effective (*Brighton Rock*) to the fluffy (*Happy is the Bride*), and from the vulgar (*I'm Alright, Jack*) to the nasty (*Twisted Nerve*). In other words, they trace the trajectory of British cinema itself.

Bill has not seen *High Treason*. An outing to the pictures organised by Camilla Betbeder would have been a perfect tonic for the troops, I say.

"We only went to see films from Eastern Europe," says Bill, dolefully.[15]

15 Bill Leader, interview with the author, February 22, 2017.

Why, if this be not education, what is?

R.L. STEVENSON, 'AN APOLOGY FOR IDLERS'

IX

A Shake to the Business

AFTER SETTLING IN AT Films of Poland, Bill paid a call to the Workers' Music Association at 17 Bishops Bridge Road, W2. Bill subscribed to the WMA newsletter, *Vox Pop* (or, in full, *Vox Pop incorporating Topic Record and LLCU News Letter*), so he knew they needed volunteers. He went along to offer his help.

"They were aware of me because I'd been selling their records and I'd started a local branch up in Bradford. There weren't that many branches of the Workers' Music Association. So I arrived and became a volunteer, helping out in the evenings, addressing envelopes, turning the handle on the duplicator to turn out the news-sheet, stuff like that. There were two paid people on the staff: the General Secretary, a person called Will Sahnow, and the clerical administrator, Camilla Betbeder. Everyone else were volunteers who came whenever they could and did the chores that needed to be done."[1]

It fell to Sahnow to devise ways to keep the WMA afloat without taking his eye off the main prize – to forge a closer link between the masses and their culture. The Topic Record Club, which sent a 78rpm

1 Bill Leader, interview with the author, January 30, 2015.

disc every month to WMA members, was Sahnow's idea. He oversaw the production and made many of the arrangements. Bill singles out TRC 10, 'Salute to Life', a song by Shostakovich, as "a really good tune". The hopes, headaches and ecstasies of a General Secretary are laid bare in 'From the Office', the column Sahnow wrote in *Vox Pop*. Here is Sahnow urging construction workers to form singing groups. The Plessey Aircraft anthem, written by workers at the Plessey factory, nursed his dreams of a singing nation. He had something of the energy of the zealot: directing the WMA Singers in a broadcast of partisan songs to Yugoslavia in 1944; reconstituting the WMA as a co-operative; undertaking missions to the South Wales coalfields to forge links between Welsh choirs and the WMA – Rhondda Unity! He was open about the scale of the task before him and ready to admit the possibility of failure. Indeed, Sahnow was so immersed in music that his Annual Report would unconsciously trace the contours of a Beethoven symphony, starting with rumblings in the low register ("We are faced with a rather shattering loss…"), modulating to a brighter key ("inevitable at a period of consolidation") and delivering a sweeping finale ("Give us the means, and we will deliver the goods!").

"He was a smallish man," says Bill. "Smoked cigarettes through a long cigarette-holder, which always looked like a bit of an affectation. His life was music and politics and organising things to happen in those areas."[2]

In issue no. 1 of *Vox Pop* (January 1942) we find Sahnow smarting at the effrontery of the big labels. Topic has been compelled to license records to Decca, who can afford to sell them at a cheaper price…

> The commercial company are issuing this and others of similar titles to ours at the price of 2/5½d.[3] thus putting a spoke in the wheels of the Club. We must rely on our members backing the Club in spite of this threat to its existence. Things won't always be thus. A time may come… And, anyway, we have other plans up our sleeve. It is possible now for us to put our hands on some pressing plant which would enable us to turn out just what we want without

2 Bill Leader, interview with the author, March 16, 2020.

3 Bill in an annotation, comments, "2/5½d. could look like an obscure masonic symbol to some folks these days. Perhaps an explanatory footnote?" Anything to oblige. *Two shillings, five and a half pence.*

> interference from commercial interests. We are investigating the possibilities of pressing our own records. There are snags – raw materials is one of them. Another is the cost of the plant! We can find comrades to work the press for us if we can get it and find raw material… There is sure to be a call on our members for share capital! So put a bit by![4]

Oh, and could members return the 10" square packaging for recycling? Alas, the situation had deteriorated further by the time of the next bulletin, a month later.

> The provision of only one record since last Summer has caused interest in the Club virtually to disappear and it is now essential to build up again. All old members are being circularised and the need for new members is as great as ever.
>
> One snag – ANOTHER RISE IN PRICE! – We now have to pay 7½d more for each record made and need to raise the price of Topic Records to 3/– to enable us to stand this new advance. Since the price of all commercial records has gone up Topics will not cost more proportionately. We can promise a superb series of Soviet recordings by way of recompense.[5]

Sahnow had the dogged stoicism of civilians during wartime, and something more – the inner certainty of a man who has been to the mountaintop. When the axe fell and the Topic Record Club was compelled to cease trading because of wartime restrictions on shellac allocation, Sahnow remained stubbornly optimistic. He wrote in the 1943 WMA Office Report:

> Let us also bear in mind the war-time restrictions we have had to face, the constant call-up of our active members for national service, the difficulty of getting paper supply for our publications, our music printers' inability to meet our demands for speedy publication of topical matter, the heavy incidence of purchase tax on our production of records and lack of raw material, the long hours of work and difficulties of travel as these have affected the personnel of our affiliated amateur music groups. In the

4 Will Sahnow, 'Topic Records Club', *Vox Pop* no. 1, January 1942.

5 Will Sahnow, 'Topic Records Club Bulletin No. 18', *Vox Pop* no. 2, February 1942.

> face of all these restrictions, necessary as we know them to be for the better progression of the war, our progress is the more remarkable. At the same time we must realise that we are only on the fringe of possibilities.[6]

With no more records to make, the WMA fell back on books.

> At the end of last year it was decided to begin the publication of small text books in a readable style in order to popularise interest in different aspects of music from the sociological standpoint. Under the title of 'Keynote' Books, the first of the series appeared in November – *Twenty Soviet Composers* by Rena Moisenco. So far 4,000 copies have been sold. The second followed in March – *Background of the Blues*, by Iain Lang. This book had an extraordinarily good sale, amounting to date to over 12,000 and has been described in the press as the best book of its kind in this country. It has brought the Association to the notice of thousands of new people, and we have received many requests for more publications on the subject of Jazz. The third Keynote, a revised reprint of Siegmeister's *Music and Society*, is just out and this will be followed up with a popular study of English Folk Music by A.L. Lloyd.[7]

This last, *The Singing Englishman*, was the first book to give a Marxist slant on traditional music. It presented the music in an unfamiliar light – as a weapon of the exploited against their oppressors. Bert's book, in modern parlance, was a game changer. I say book; *The Singing Englishman*, like the other Keynotes, was closer to a pamphlet. Nevertheless, its impact was great – not least on its publisher!

Come peacetime, the Topic Record Club renewed its activities in 1947 with TRC17, 'Soviet Airman's Song' b/w 'Chapayev' by baritone singer John Hargreaves, with piano accompaniment by Arnold Goldsbrough (mark the name). By 1950, it was up to its regular monthly rate.[8] The first Topic Record Club release of British folk origin came the same

6 Will Sahnow, 'Report of Workers' Music Association on the period July, 1941–June 1943', *Vox Pop* no. 2, November 1943.

7 Ibid.

8 Though twelve records a year would be rather too neat. The Complete Topic Records Discography (from *Three Score & Ten*; TOPIC70, 2009) lists 15 releases in 1950 and 15 releases in 1951.

year. TRC 39, by Ewan MacColl, has two songs a side – 'The Asphalter's Song' and 'I'm Champion at Keeping 'Em Rolling' b/w 'Fourpence a Day' and 'Barnyards of Delgatie'. The generous helping possibly reflects uncertainty of the market. Yet, following Bert's book, workers' music was now officially synonymous with folk music. The very next Topic release, TRC40, 'The Four Loom Weaver' b/w 'McKafferty', also by MacColl, consolidated the direction. As for songbooks, *The Pioneer Songbook* celebrated the centenary of the Co-op movement, and was in the old community singing mode, but the irresistible tide swept onwards with Patrick Galvin's *Irish Songs of Resistance*, a history of Ireland as reflected through her songs, a MacColl selection, *Scotland Sings*, and *Coaldust Ballads*. The latter offered Sahnow arrangements of North East mining songs sourced from Bert Lloyd's *Come All Ye Bold Miners*. The clincher came with MacColl's *Shuttle and Cage* (1954). Its thesis – that folk jumped from rural to urban during the industrial revolution – constituted a poke in the eye for the maypole brigade.

The task of notating all the songs fell to – who else? – Will Sahnow. Mindful of the WMA's commitment to choral groups, he painstakingly devised a piano accompaniment, made arrangements for four-part harmony and generally turned rough music into something suitable for the parlour.

"We were living in the old world of tenor, bass, soprano, contralto. The WMA was pursuing the choir, as a collective, as an organisation – a way of keeping people together. And if you've got a bunch of people interested in music you have to have something for them to sing. It was music for the drawing-room, in a way. It was the Peter Pears and Benjamin Britten view of folk songs – piano accompaniment, standing at the piano…"[9]

Sahnow would fit the words below the stave with a piece of office kit called the Vari-Typer.

"It was really a big old clunky typewriter," explains Bill, "but instead of keys coming up and hitting the ribbon, the Vari-Typer used a type shuttle that was interchangeable. You took it off and put another one on, which gave you another typeface. It had a huge selection of typefaces, every one as ugly as sin, as you can see here." (Bill gestures at my copy of *Popular Soviet Songs*, handily lying about.) "The best ones are pretty ugly, and the bad ones are really horrible."

9 Bill Leader, interview with the author, October 5, 2012.

Yet Will Sahnow's Vari-Typer, he says, "saved us a bloody fortune."[10]

When I first became aware there was such a thing as a WMA archive, it was tied with pink string and scattered in cardboard boxes under John Jordan's bed.[11] John Jordan? Bill waxes vague when I ask about him:

"He was newly in as I was heading out. He was younger than me, but then most people are. I didn't realise he'd got some sort of position."[12]

What happened next was that Jordan died, and Glasgow University obtained the WMA archive for its Political Song Collection, to augment Janey Buchan's extensive archive. As an academic resource, the archive is more accessible in a university than under someone's bed, although Glasgow is still a long way from Manchester. Thus, shortly before the first lockdown, I find myself on Glasgow University campus perusing yellowing papers of hypnotic interest.[13] John Powles, the keeper of the Collection, is a courteous and dignified host. It will take a longer stay for me to do full justice to the documents, but I do my best in the limited time available, before the office closes and I'm due to take a return coach.

Music and Life, a quarterly edited by Alf Corum, is your standard crude, cheaply produced newsletter, of a piece with the WMA *Bulletin* (I shall come to the *Bulletin*, which justifies the Glasgow trip alone): four sheets rattled off on a typewriter, hastily duplicated, casually folded, and, on a good day, stapled. *Music and Life*, from 'the music group of the Communist Party of Great Britain', promulgates the party line on matters musical. Its contributors' opinions about music are as unapologetic as their opinions about politics, and inconsistency is an ongoing risk. No-one can bring themselves to ask if opera sung in English is compatible with the internationalist spirit of marxism.

The party line is that rock 'n' roll is a degraded version of the blues, but the music group of the CPGB reserve their special ire for skiffle, which, as well as being a travesty of a genuine folk art, threatens the

10 Bill Leader, interview with the author, January 30, 2015.

11 A Mudcat thread from 2004 about the song 'I Pity the Poor Landlord'.

12 Bill Leader, interview with the author, June 15, 2018.

13 The present tense in this case is March 3, 2020.

livelihoods of MU members. Keep Music Live and Keep it Punctilious! There is no substitute for live music, except perhaps in Prague.

I'm drawn to the headline 'Where Records are No Menace to Musicians' in a 1959 copy of *Music and Life*. MU Secretary Harry Francis, lately back from the Prague Spring International Festival of Music, praises the Prague Theatre of Music, where music lovers can enjoy record recitals in cosy armchairs, to the gentle accompaniment of tasteful light projections.

> Youth audiences are taught the difference as between the best in jazz, and the gimmick-ridden rock 'n' roll type of rubbish of which we in the West have to suffer so much these days. The best international artists in the field of jazz are well-known to Czechoslovak audiences.[14]

Don't tell Harry, but gimmick-ridden metal prevails in the Czech Republic these days. The surprising Prague Spring reference suggests the phrase was current a decade before the events of January-August 1968.

Also in *Music and Life*, John Vyse excoriates Duke Ellington for pandering to "the decadent pseudo-primitive cult of American intellectuals"[15] and diluting the lifeblood of a people's music with his "cheap exhibitionism". Notwithstanding, he turns up as the jazz correspondent of the *Daily Worker*, but it seems folk is also in his brief, because his column on July 9 1954 is devoted to a round-up of Topic 78s. Vyse captures Topic in transition at precisely the moment Bill and Alex Eaton were advancing the WMA cause in Bradford.

The selection is headed by TRC 67, 'The Bonny Boy' b/w 'She Moved Through the Fair' by Patrick Galvin. Vyse resorts to vague blandishment – "Patrick Galvin is the latest singer to be added to Topic's growing list of stars." Galvin, a good poet and an indifferent singer (in Bill's opinion), was also responsible for the Keynote *Irish Songs of Resistance*, which quickly ran to a second print. Topic TRC 65 features Sadlers Wells opera star Anna Pollack on two soldier ballads – 'Mrs McGrath' (from the 'Johnny I Hardly Knew You' school), and 'Shule Agra'. Alan Bush conducted the former and Mátyás Seiber provided an arrangement for the latter.

---

14 Harry Francis, 'Where Records are No Menace to Musicians', *Music & Life* no. 10, Summer 1960.

15 John Vyse, 'Jungle Bands and Jitterbugs', *Vox Pop* no.6, December 1942.

"Mátyás Seiber of course was a Hungarian composer who wrote classical music but also arranged things. That record is the last throw of the old school," says Bill.

Never again could opera singers sing folk songs with impunity, and Laurence Olivier would rue the day he dared to play Othello in blackface. It was the end of manifest destiny and the beginning of self-determination, a hard struggle. Vyse, catching the whiff of change, comments, "'Mrs McGrath' might sound better if the singer had had two or three pints." Tellingly, three Chinese folk songs by soprano Yu I-Hsuan on TRC 66, are "lovely examples of modern Chinese art arrangement *and very easy to follow for anyone familiar with British folk music*" (my emphasis).[16]

TRC 64 presents two wartime Polish ghetto songs, 'Wilno Ghetto Song' and 'Warsaw Ghetto Song', by Martin Lawrence with the WMA Singers conducted by Alan Bush. "These two most moving laments are sung with delicate sympathy in Yiddish and English. The WMA Singers are supreme at this sort of thing." Bill remembers the singer. "Martin Lawrence was a cantor from Stepney, but he sang opera when he could get the chance." His sacred singing can be heard in the synagogue sequence of *Hand in Hand*. The 1961 film proposes reconciliation between the two old religions, Judaism and Christianity, as represented by two innocents, more accurately described as blood brothers than childhood sweethearts. It would be impossible to remake *Hand in Hand* for a lot of reasons, most tending to show how progress is an illusion.

'Ballad for Americans' by Martin Lawrence and the London Youth Choir (TRC 61-2) is a "modern declaration of progressive America," writes Vyse. "Earl Robinson's classic will retain its popularity for many years to come," he predicts.

"That didn't happen," says Bill, before conceding that Earl Robinson was a major tunesmith.

Folk song in the Eisenhower era went about in heavy disguise. 'A Knave is a Knave' became, in Doris Day's hands, 'A Guy is a Guy'. An old Armenian tune tethered to coy sexual promise resulted in 'Come on-a My House', a hit for Rosemary Clooney in 1951. 'Let's Walk That-A-Way', a 1953 hit by Doris Day and Johnnie Ray, comes from 'Ha

16 'Best Sung on Two Pints', John Vyse, *Daily Worker*, July 9, 1954.

Ha, This-A-Way', a song by the ex-convict responsible for the Weavers' two-million selling 'Goodnight, Irene' in 1950. The ex-convict, of course, was Huddle Ledbetter. Guy Mitchell sang *She had a dark and roving eye, and her hair hung down in ringlets...*, and charmed listeners with his raffishness. 'The Roving Kind' sold very well in 1951 "without many people realising it was about venereal disease," says Bill.[17] It was left to 'Tom Dooley', a wholesome murder ballad from the collegiate Kingston Trio, to open the folk floodgates in 1958.

Something was happening. A *Music & Life* editorial detected the signs of dialectical materialism: "Marxists know the significance of those movements in Society which have the seeds of growth in them as against those, apparently more important, which, internally, are in decay."[18] The trouble was, WMA founder Alan Bush, General Secretary Will Sahnow, clerical administrator Camilla Betbeder and chief technical officer Dick Swettenham alike professed ignorance of the record market. One man knew. The volunteer from Bradford knew. In late 1956, Bill was made the production manager of Topic Records, a newly created post. He was given the stipend of £10 a week, which, being equal to his wage at Films of Poland, emboldened him to permanently exchange a small corner of the film world for a small corner of the music world.

A change now came over the WMA. It can be detected in the pages of *The Bulletin*, the WMA newsletter formerly called *Vox Pop*. A standard appeal for volunteers, as penned by Will Sahnow, might read: "We appeal to any readers living convenient to our offices who are free of other daytime duties to come forward and offer their services, either voluntarily or in a paid capacity. A shorthand-typist in particular is urgently needed..."[19]

Severe, efficient, with only a hint of anxiety. Suddenly we find...

> The TOPIC RECORD COMPANY offers excellent positions to people interested in its fascinating and flourishing work. Due to its expanded recording schedule and the opening of new recording studios, it now has vacancies for secretarial workers, aspiring journalists, commercial artists, carpenters, painters and decorators, electricians, and anybody interested in records and workers' music.

17 Bill Leader, interview with the author, April 4, 2012.

18 *Music & Life* no. 5, Summer 1958.

19 *Vox Pop*, May 1943.

> Renumeration: the satisfaction of a good job well done. Applications (in overalls please) to 17, Bishops Bridge Road, any time up to 10.45 p.m. and from then until 11.00 p.m. at the Prince of Wales saloon bar.[20]

One can sense the careless shrug if no-one came forward. The streak of drollery, that got Bill into trouble on the soapbox on Broadway, is to the fore. The launch of stereo in 1958 (it happened earlier in the USA) is a gift to the inner satirist:

> I HAVE JUST HAD the distinguished pleasure of listening to the new 50-50 stereo disc system which has been, as it were, perfected by the engineers of the Spectra group of labels and which will be used on the Bawlophone and Quaver labels towards the end of this year.
>
> Both these labels have a large output of popular music, and as their artists manager Ned Sidelman said to me at the cocktail party with which these amazing records were launched, "This has put new life into the gimmick business."
>
> I sat enchanted throughout the whole of the new LP entitled *Stereo for Young Lovers*, and marvelled at the amazing realism with which the singer's voice appeared to alternate swiftly from one loudspeaker to the other, changing places with the instrumental accompaniment, which was busy alternating in the other phase, as it were. In the number 'Lonely Wanderer', a very sad ballad indeed, the singer appears to wander all over the room, and only comes to rest inside the left-hand speaker at the very end of the song. In the poignant verse in which he says that he is down at the dockside looking for his love, a very touching ship's siren hoots softly towards the back of the room, and in the following verse where he wanders by the railroad track, an extremely realistic train sweeps past from right to left. Altogether a wonderful disc.[21]

Bill makes a very, very good comic writer and might have matured into P.G. Wodehouse or at least Chris Welch,[22] if sound engineering hadn't

20 'Situations Vacant', *Bulletin*, April 1957 (it confusingly says March on the cover).

21 Bill Leader (uncredited), *The Bulletin*, October 1958.

22 Chris Welch, a music scribe and amateur drummer, big on progressive rock, wrote very irreverent and surreal singles reviews for *Melody Maker* in the seventies.

come first. Music industry insider Ned Sidelman is his great creation. Ned first appears in the April 1957 edition of *The Bulletin*, where he is introduced as 'Topic's top talent scout'.

"*He is the boy who first introduced The City Ramblers to the record-buying public,*" I recite to Bill, fresh from Glasgow. Bill's chuckles are muted, I would say. I study his features for a sign of wistful nostalgia and return empty-handed. "*The Ramblers' record TRC 101 is still available and selling like hot petrol*! Is this your stuff?"

"It is, yeah. Yeah."

"*The Ramblers themselves are now several rungs further up the ladder of fame than when we issued their recording. They are touring the Halls with tub bass and washboard, pausing only to televise and broadcast. Another of Ned's discoveries is Nancy Whiskey. An LP of this petite Scots singer is in preparation and has been for some considerable time.*"

"That's true."

"That is, you recorded it, but it sat on the shelf awaiting release?"

"Well money wasn't readily available for the production of records, and especially for the production of records that we suspected wouldn't sell very much, although we didn't require them to sell very much. It was very easy to choose other things. I forget all the detailed reasons why Nancy Whiskey didn't get a ready release."

"But then she was on the top of the hit parade, yeah?"

"Yes. Well I think it was the top, wasn't it? It stayed there for a bit."

('Freight Train' by Nancy Whiskey with the Chas McDevitt Skiffle Group reached the top five in the UK Singles Chart in 1957.)

"That must have been a bit of an impetus at least. Or not?"

"Yeah, well one would think so. We weren't corrupted by such considerations."[23]

No indeed. The aim was a people's culture untrammelled by commercialism or vested interests.

The WMA Summer School is Will Sahnow's enduring legacy. The WMA's Annual Report of 1957, a grim read, is relieved by a single line, "A gross surplus of £87 was made by the School."[24]

---

23 Bill Leader, interview with the author, July 11, 2019.

24 From the WMA Archive, housed in the Political Song Collection, Glasgow University.

"The first WMA Summer School I went to was in 1954 at Wortley Hall, between Sheffield and Barnsley, and I got there easily, because I was coming from Bradford, which is just up the road. I think it was possibly the first WMA School at Wortley Hall.[25] They went on for many, many years at Wortley Hall. I went to many Schools over the years, and they were always exceptionally successful. They were the only event at that time where people who were interested in folk music, choral singing, symphonic music, chamber music, brass bands and jazz were all locked into one place for a week to get on with it. That was quite a thing. So they were successful events, culturally and socially and sexually."[26]

(Yes, by the late fifties the dormitories of old had been replaced by small, individual rooms with their own washbasins.)

Lecturers at the 1956 School included composer Alan Bush, brass band composer Alfred Ashpole, experimental physicist and skiffler-in-chief John Hasted, and organist/conductor Arnold Goldsbrough, the accompanist on the B-side of TRC 17.

"An amazing bloke. I remember him sitting down to play something at the piano, which they'd actually got tuned for the School. It was a grand piano that was not all that grand, because it had been tuned down a semitone, because it was feeling its age, and you don't want to demand too much out of your grand piano. Goldsbrough didn't know about this until he sat down to play, and he saw the music and he pressed the right levers, and the wrong sound came out. So he then continued to play, transposing at sight everything up a semitone. He was a good working musician, and a good conductor."[27] Also present were Philip Hecht – "he had been, but I think was no longer, sub-principal of the Hallé Orchestra" – and singer Felicity Bolton, who lived on the top floor of Bishops Bridge Road, and gets a mention in Bill's report of the Summer School in *The Bulletin*:

---

25 The inaugural 'WMA Holiday School of Music' was held in Edinburgh in 1947, under the direction of Rutland Boughton. A fascinating character, Boughton wrote a very successful opera, *The Immortal Hour*, and attempted to turn Glastonbury into a kind of Bayreuth, organising the first Glastonbury Festival in 1914, more than half a century before Michael Eavis.

26 Bill Leader, interview with the author, March 16, 2020. Cf. Stan Kelly: "There was always a lot of shagging going on at Wortley Hall, wasn't there?" Oswestry, May 19, 2013.

27 Bill Leader, interview with the author, March 16, 2020.

The WMA Summer School, Wortley Hall: Owen Bryce with trombonist

John Hasted takes a banjo lesson

Hasted stoops over a lady with a guitar (collection: Bill Leader)

A Great Day at Wortley Hall. Teachers and alumni of the WMA Summer School, 1956. That's Bill and Gloria (semi-obscured), far left, second row standing, and Alex and Louise Eaton, characteristically radiant, below the man in the white t-shirt astride the Ionic column, right; Wendy Corum, hand resting on the jutting stone, is raised on the Ionic column, left. We score highest for identification with the front seated row: unknown, Will Sahnow, Arnold Goldsbrough, Felicity Bolton, Alan Bush, Philip Hecht, unknown (collection: Bill Leader)

> Down this passage to a room with a good ring to it. This is to flatter the voices in the solo singing class. Tummy in. Head well up. One alone to be my own. Fairest isle, all isles excelling. Over the past four years Felicity Bolton, who has taken this class, has developed a technique in instructing large classes in the rudiments of breath control and posture, thus leaving more time for dealing with individual coaching. The class is called the John Goss Memorial Course, in memory of that great British baritone, who did so much for Workers' music.[28]

Bruce Turner and Owen Bryce took care of the jazz. The Summer School Seven, under the direction of Owen Bryce, filled Wortley's ballroom. Bryce, one of the founders of traditionalist jazz in Britain, "had a Cyrano de Bergerac nose, quite a hooter," recollects Bill. "That's why he played the trumpet, I think, because he had plenty of room for it."[29]

> In the other wing a middle-aged lady who is a dab hand on the glockenspiel, but finds herself a little short of material, is persuading a young man who is writing a sonata for Northumberland pipes to adapt it to her instrument.

"Everybody played or got up and sang or did something on the last night. There were choral things, there was jazz, there was the folkie people and so on. They also had this performance of 'The Ballad of Wortley Hall.'"

On its first airing in 1954, Bill's ballad about the transfer of Wortley Hall from the ruling class to the workers caused the Earl of Wharncliffe to walk out (but I think I've told you this).

> On the final night all hair is let down. All hell is let loose. The student conductors conduct the choir. The star vocal students let off their fireworks. That glockenspiel sonata gets its world premiere. And you, although you thought that you never would, find yourself performing that musical feat that had been your secret ambition.

---

28 Bill Leader, 'Summer School Special', *The Bulletin*, March, 1958, and the following extracts.

29 Bill Leader, interview with the author, July 31, 2019.

Meanwhile Will Sahnow was destroying his health for the sake of music and liberty. He ran a music course at an adult education centre in Hendon on evenings; "so he never stopped," says Bill. On November 17, 1956, he was the guest of the Co-operative Society in Warrington, at a conference headed 'Music in Co-operative Education'. His message was the familiar one: musicians need to work towards a common goal. The 21st anniversary of the WMA in the coming year, 1957, was to be marked with celebratory conferences and concerts. Sahnow threw himself into their organisation with his usual zeal and energy. He died in his sleep during the night of January 15, 1957. The cause was given as heart disease, exacerbated by overwork. He was 59 years old.

Gerry Sharp was appointed Sahnow's successor as General Secretary. His background was in administration and accountancy: he had previously balanced the books at Unity Theatre. Sharp continued Sahnow's 'hands across the ocean' policy. This raises the vexed question of Anglo-American relations as they pertained to the music industry in the mid-fifties. What was that line about "two nations separated by a common language" (© George Bernard Shaw)? Contracts were involved, to raise the stakes a little.

Every schoolboy knows that Americans have the most advanced Capitalist system in the world, which fits with constitutionally prescribed notions of 'the pursuit of happiness' and justifies any amount of misery in citizens on a low income or nations unfortunate enough to occupy its backyard. Whereas in England (this applied up to the mid-fifties), there was no pleasure beyond the satisfaction of doing one's duty. In Topic's terms, this meant making the world a better place through folk music. The notion was quaint or outright naive to your typical hard-nosed American businessman, but, after all, folk music was a commodity like any other. As it happened, Topic got along better with the rapacious buccaneers of Riverside Records of West 51st Street NY, with their jazz and fast cars, than scholarly, esoteric Folkways Records of West 46th Street NY.

Some background about the state of folk music in Britain in the fifties might be useful here. As far as mainstream media goes, traditional music in Britain was confined to *As I Roved Out*, a half-hour broadcast on Sunday mornings on the BBC Light Programme. Reg Hall, an important figure in Bill's life, affects the fruity tones of a period BBC announcer to convey its flavour:

"'*It was fortuitous that we actually found ourselves in Norfolk three weeks ago, and, by luck, it happens to be the village where Harry Cox lives. Harry, what would you like to sing for us?*'"

Reg dismisses *As I Roved Out* as "a travelogue with incidental bits of music in-between."[30] Gloria, on the other hand, enjoyed its friendly way of introducing the old songs to a wide public:

"We first heard folk music on Sunday morning, *As I Roved Out*, with Peter Kennedy. It was very good."[31]

The major labels ignored folk song; Bill maintains, "HMV produced a 78 of Ewan MacColl, a 78 of Bert Lloyd and a 78 of Isla Cameron. Otherwise, there was nothing."[32] He dismisses the Princess Margaret-inspired barndance craze of the early fifties as a flash in the pan.

This pleasant stagnation might have continued indefinitely, were it not for events in North America. The demand for novelty, the diversity of the world's greatest melting pot and a constitutionally prescribed commitment to the pleasure principle resulted in an outbreak of hi-fi mania. Often this took an exotic manifestation. Belgian steam organ vied for attention with self-help belly-dancing (*How to Make Your Husband a Sultan* by Özel Türkbas is nowadays sought for its musical quality). Mushrooms (*Mushroom Ceremony of the Mazatec Indians of Mexico*, on Folkways) and frogs in audio fidelity (*Sounds of North American Frogs*, ditto) jostled in this incredibly strange marketplace. It was around this time that Kenneth S. Goldstein, a business graduate from the City College of New York, wondered if English folk music might have a place amidst such fabulous diversity. He struck a deal with Stinson Records and approached the WMA to find suitable material.

The parallels between Stinson and Topic are striking. Stinson was founded in 1939, the same year as Topic, with the same Soviet bent: the label sprang from the alliance between Herbert Harris, an entrepreneur with a record shop and a cinema devoted to Soviet film among his assets, and Irving Prosky, an emigre who distributed Soviet

30 Reg Hall, interview with the author, April 18, 2016.

31 Gloria Dallas, interview with the author, July 20, 2016.

32 Bill Leader, interview with the author, April 4, 2012. Alistair Banfield, that indefatigable folk discographer, points out that HMV produced *two* 78s by Ewan MacColl and *two* 78s by Bert Lloyd, whilst confirming that Isla Cameron did have one 78 on the label. Otherwise, there was nothing.

recordings in the USA. Stinson championed socially aware singers, exemplified by Woody Guthrie.[33] When the deal fell through because of Stinson's ineptitude, Goldstein approached Riverside with the same arrangement. Riverside, of course, is primarily associated with modern jazz.

"The Loneliest Monk, or whatever his name was, Philly Joe Jones and all that lot, and racing cars. They were into car racing records. *Vroom! Vroom!* I might even have that record, *The Grand Prix of Gibraltar* somewhere."

The Grand Prix of Gibraltar by Peter Ustinov, RLP 12-833, at once celebrates and mocks the label's fondness for car racing discs. Bill owns a copy because Riverside were in the habit of sending vinyl in lieu of royalties, and sports car discs were more disposable than records by Thelonious Monk, Sonny Rollins and Bill Evans.

A contract was inked between Topic and Riverside with Kenneth Goldstein as mediator. There was one snag: WMA had no recording facilities. Happily, an enterprise called Magnegraph, located on Hanway Street, W1, hired out recording equipment of a professional standard. The task of recording the sessions fell to Topic's new production manager.

The deal with Folkways came later and was supposed to be mutually advantageous. Topic would release Folkways material in the UK and Folkways would reciprocate by issuing Topic material in the States: basically, we'll swap you your Woody Guthrie for our Ewan MacColl! Folkways, formed by Moses Asch in 1948, without being the eerie mirror-image of Stinson, pursued a similar business model: 1) less is more 2) keep expenses to a minimum 3) defer all payments for as long as possible 4) keep titles in print despite meagre sales and 5) avoid having a hit at all cost.

"Moses Asch built up a huge and fantastic record label on nothing really. He recorded all these rare recordings…all based on academics wanting to get something published out of their research. They got something they could claim as a publication, and he got a record. He had to *make* the record, of course."[34]

Moses, or Moe, was the son of Sholem Asch, a Yiddish writer described as "heavy-set, stoop shouldered and amiable". A family

33 See https://en.wikipedia.org/wiki/Stinson_Records.

34 Bill Leader, interview with the author, April 4, 2012.

friend observed that Moe "walks, talks, stoops and shouts just the way the old man did."[35] Sholem set his short stories and novels in the vanished world of the shtetl, the setting of Moe's Polish childhood. This background explains Moe's interest in fungi. As he tells it:

> As an infant in Poland we spent our summers searching for strawberries and mushrooms. This has stood me well these days for I have been able to discuss the mushrooms of Poland with ethnomycologists such as Mr. Wasson, for whom I issued *Ceremonies of a Mushroom Worshipper.*[36]

And somewhere in Moe's office on West 46th Street lurked a Vari-Typer! The outlandish typewriter determined the look of booklets inserted into the pockets of Folkways' thick cardboard sleeves.

"Very sophisticated. You could change the type and change the spacing, but none of the type was designed to be read. And then Asch used offset litho in its early days, so it was all soot and whitewash reproductions. It was all pretty horrendous."

The thaw in the relationship started when Topic omitted to pay the agreed $50 per recording licensing fee. This was a book-keeping oversight, claimed Gerry Sharp and rectified the matter. Then Topic sent royalty payments without deducting UK tax, leading to a double tax headache for Folkways' accountant.

As part of the arrangement, Topic's production manager was tasked to choose records for UK release from selected Folkways titles. This is how the Topic catalogue came to rejoice in *The Music of New Orleans Volume 1* (12T53, 1959) and *The Music of New Orleans Volume 3* (12T55, 1959), but not, frustratingly, *The Music of New Orleans Volume 2.*

"They had three records of live recordings made in New Orleans by Sam Charters. But two were available and the third one had already been taken up by Melodisc. They issued it."

---

35 'Folkways on Record' by Robert Shelton, *Sing*, vol. 5, no. 4 (August 1951), p. 69.

36 'Moses Asch Speaks Out! Folk Music – A Personal Statement', *Sing Out!* (February-March 1961), cit. by http://www.radiohazak.com/Asch.html. The disc in question, *Mushroom Ceremony of the Mazatec Indians of Mexico*, from 1956, is a remarkable document of a religious ceremony centred around psychedelic mushroom use. The description on the sleeve of FR 8975 is mouth-watering: "Dr Wasson is able to prove that the words and undoubtedly the chant are pre-Cortes, going back for many centuries… it is rivalled in the Old World only by the Vedic chants of India… Recorded… in the Mazatec Mountains in the northern corner of the State of Oaxaca, July 21, 1956."

Yes, Topic got *The Music of the Streets* and *Music of the Dance Halls*. *Music Of The Eureka Brass Band* (Melodisc MLP 12-110) got away. Bill still beats himself up about this.

"Eureka was one of the famous brass bands, street bands, and of course that one sold hundreds, if not thousands, to the enthusiasts here. Ours didn't sell very many, because they were more difficult than the brass band thing. They got in ahead of us."

And then there was the case of *Spirituals* by the Fisk Jubilee Singers (12T39, 1958). Samuel Charters tells the fascinating story of the Fisk Singers on RBF Records RF 5, *An Introduction to Gospel Song*.[37] I shall attempt a summary here.

In October 1871 eleven students from Fisk, a black university in Nashville, Tennessee, set out with the school treasurer and a woman chaperone on a tour to raise funds and awareness. The first stop was Cincinnati, Ohio. The small group would sing in churches in the northern states, spreading the university's name whilst hopefully boosting its coffers. The tour, which was expected to take two or three months, in fact lasted for three years, and encompassed the United States, Europe and Great Britain. The Fisk Jubilees performed in front of President Ulysses S. Grant and various crowned heads, including Queen Victoria, and set Fisk University on track to becoming the prestigious seat of learning it is today. More, the gospel choir demonstrated the artistry and pride of newly freed African Americans.

"A really tedious musical experience if there ever was one," says Bill.

"That bad, eh? I haven't heard it."

"Save yourself forty minutes. I can see why Folkways did it, because musicologically they're of significance to America, but there's no reason for issuing it here in this country. I don't know why I went for it. But I did."[38]

Yes, trying to downplay our musical heroes' recent lack of form, we hastily say, "They used to be great. You should have heard them in 1871." Yet the sleeve, a wonderful design from Gloria, would tip the balance in favour of 12T39, should I chance upon a copy. No excuses are needed for 12T29, *Brownie McGhee and Sonny Terry*, and especially not for *Harmonica Blues* by Sonny Terry (10T30, later 12T30).

---

37 RBF Records – it stands for Record, Book and Film Sales, Inc. – is a subsidiary of Folkways, and shares the distinctive company look.

38 Bill Leader, interview with the author, March 16, 2020.

"They're classic. Particularly Sonny Terry. A brilliant piece of work. That was OK. Then I picked the arse-end of the *Music of New Orleans* thing, and missed the one that was popular, and then chose, for some reason that I ought to have my head examined to find out why, the Fisk Jubilee Singers. Oh, we did the Pete Seeger guitar thing, and his banjo tutor."

The reference is to *Pete Seeger's Play the 5 String Banjo*, which crossed the Atlantic and changed from FI 8303 to 10T23, prompting this letter to Bishops Bridge Road from the great man.

Pete Seeger
Dutchess Junction
Beacon, N.Y.

August 15, 1959

Dear Friends,

Just a belated note of thanks for your sending me a copy of my record (the banjo-instruction record, I mean). The brochure was beautifully put together and a great improvement on the American edition, I feel. However, in my particular copy, there was a misprint and two of the pages were blank. I wonder if you have another copy which you would be able to send me.[39]

"Very good. Very good. 'Thank you for the crappy production,'" chortles Bill.

The banjo tutor had been preceded by a guitar tutor, *Pete Seeger's Guitar Guide for Folk Singers*, 12T20, 1958.

"That hit at about the right time. People were beginning to realise you could do more with a guitar than just three chord strums. I mean, although Folkways were much more successful in what they did than we were, there was a sort of comparable thing going on. It was a very short-lived thing, and why it finished I can't remember."[40]

39 From the WMA Archive, housed in the Political Song Collection, Glasgow University.

40 Bill Leader, interview with the author, July 11, 2019.

It had something to do with Mose Asch getting annoyed when Topic began to discount Folkways titles (this around 1960). He remonstrated with Gerry Sharp that the integrity of his label was being compromised. Topic's General Secretary replied that this was standard trade practice and anyway, "I understand that this is what you did with the Topic records you had in stock when you wanted to clear them."[41]

A meet-up on one of Moe's regular visits to London would doubtless have cleared the air, but no such meeting took place.

"Gerry was quite good at thinking up reasons why not to pay people. And Moe was pretty smart. Two pretty smart people. Though I think Moe was probably, in the end, the smarter person. But they were a good match," summarises Bill.

The alliance ended with a whimper rather than a bang. The last Folkways to appear on Topic dribbled out in 1965 – *More R & B From S & B* by Sonny Terry and Brownie McGhee (TOP 124), which was also, coincidentally, the last EP released by Topic.

"Hmm! You avoided all the really out-there things Folkways had."

"What, *The Rainmaking Dance of the Flat Head Indians of South-East Patagonia*?" shoots back Bill, showing that the spirit of Ned Sidelman is alive and well.

For Folkways, international folk was not a commodity so much as a source of anthropological study. This approach created vitally important documents. *Jamaican Cult Music* (P 461) captured roots reggae in embryo. *Islamic Liturgy: Song & Dance at a Meeting of Dervishes* (FR 8943) introduced western listeners to the concept of trance as entertainment. A box set, *Negro Folk Music of Africa and America* (P500), presented enough material for listeners to construct their own musical genealogy of the black diaspora.

"Where he was selling, of course, was to the education market there. Every university, of which they had many, had a library, and some of the better schools had that sort of library, so he was selling to the education market, and then to the enthusiasts, and he was getting it for nothing, I think. People would use him as part of publishing their scholarly work."

"It would be a line in their CV."

"Yes. He didn't have to pay them very much."

---

41 Tony Olmsted, *Folkways Records: Moses Asch and His Encyclopaedia of Sound* (Routledge, 2003), pp. 98-99.

The scholarly rigour applied to, say, *Traditional Folk Songs of Japan* (FE 4534) came with an eccentric occult tinge when applied to the homegrown product. *The Anthology of American Folk Song* (FA 2951-FA 2953), a landmark three-box set, presented Depression-era music according to function (play, dance, worship and work), and was accompanied by alchemical diagrams and esoteric annotations from Harry Smith. The various genres – blues, gospel, sharecropper, hillbilly, hokum and proto-C&W – came together for the first time, and, in keeping with the wishes of Harry Smith, the series' editor and a student of the Kabbalah, they were sited in harmony with the music of the spheres. The performances, which are uniformly staggering, expose the inanity of the popular song of its time, the popular song of our time and doubtless the popular song of times to come. The majestic music of Clarence Ashley, the Rev. Gates, Cleoma Falcon, 'Dock' Boggs, Blind Willie Johnson *et al* is a gift for the ages.

"God, I should be looking for Folkways Records, not Leader and Trailer rubbish," declares Bill, in a moment of pitiless self-awareness.[42]

42 Bill Leader, interview with the author, June 15, 2012.

Being brought up in a seaside town, you'll find these underground entertainers who are absolutely honest!

TONY HANCOCK, *FILMS AND FILMING*, AUGUST 1962

X

Ducking and Diving with Punch and Judy

By 1956 the writing was on the wall for the humble 78, and the majors were battling over the next big thing. The Columbia corporation threw its weight behind the format it had introduced in 1948, the 12"/30cm, 33⅓rpm long-player, or 'LP', whilst RCA Victor backed the 7"/17.5 cm, 45rpm disc, or 'single'. The LP had greater playing time, and offered a superior package, with, at best, an attractive design and space for illuminating or diverting text. The format encouraged rapt absorption and was very good for symphonies and jazz and such. Whereas the single had a smaller diameter and a higher speed and (mostly) came in a generic, plain sleeve. It delivered energy in concentrated bursts and was well-suited for the quick adrenalin rush of pop music; it was also cheaper.

The arena of contest was the USA, simply because America boasted the most advanced record market in the world. Whereas the record industry in the UK was underdeveloped and undercapitalised, and Topic was more undercapitalised than most. Its records were aimed at a coterie who would be horrified to be mistaken for consumers. The tiny label was part of a larger operation, the Workers' Music Association, which published textbooks and songbooks. They employed *two* full-time staff. Their preoccupation was nothing so frivolous as entertainment;

the WMA proposed nothing less than societal change by means of music. They were motivated by art and idealism. Business was not even a poor third; it was an evil necessity at best. They set themselves against the uncritical values of untrammelled consumerism and rejected such capitalist constructs as the free market and the law of supply and demand. Logically they should have connived at their own destruction, but why make it easier for the capitalist interests ranged against them?[1]

The balance sheet of December 31, 1956, taken with the Annual Report of 1956, showed that the WMA would never get through 1957 without taking drastic measures.

"I don't think I was sufficiently aware of the pressures to get suicidal about them, but maybe if I'd known more, I would have been less sanguine," says Bill, with his customary sanguinity.[2]

The WMA, as good Marxists, were much preoccupied with *the means of production*, and adopted the singular format of the long-player *with an 8" diameter* from a mixture of desperation and expediency. "We learned to believe that two formats were better than one," writes Bill, but when the last 78 rolled off the line in 1958, *four* new formats vied for attention (see Fig. 4).

Records were expensive in the UK because they were classed as luxury items. The rate of Purchase Tax varied, but at times it reached 70% of the net retail price (during the war, mostly). As a sop to the music societies – because *proper* music was something to be encouraged – tax was waived on records produced in a limited quantity. This was defined as 99 copies. As tax was eligible on further pressings and, retrospectively, on the initial 99 copies, the system punished entrepreneurialism.

"We had to sell something like 300 records or so to start making a profit. And these were the days when people didn't dash out and buy records at the first opportunity. These records cost about six shillings each. So if you sold a record, the profit was probably pence."[3]

The Topic price structure breaks down thus:

1 "Most of the times that the WMA gets mentioned it ends up sounding as though its members were poe-faced dogmatists; whereas most of the members had dance, song, and an enjoyable time as their immediate aim. If they could also improve society in the course of having pleasure, that would be a wonderful bonus. Of course there were some dour humourless types too." Bill Leader, annotation, Jul 3, 2020.

2 Bill Leader, interview with the author, March 16, 2020.

3 Bill Leader, interview with the author, September 8, 2009.

	s.	d.
10" 78 rpm	6	6
8" 33⅓ long-players	16	8½
10" 33⅓ long-players	28	2
12" 33⅓ long-players	37	8
7", 45 rpm extended plays	12	10

Fig. 4 *Topic formats and prices (1956-61)*

A trawl through *Sing* back-copies is informative, if distracting. The postbag of *Sing* vol. 2 no. 2 (June-July 1955), for instance, contains just two letters. We know both the correspondents. Alex Eaton of Shipley, Yorks, praises *Sing*'s song content as invaluable "to struggling provincial groups" and suggests more traditional material to inspire contemporary songsmiths. John Foreman of London NW5 requests material suitable for traditional jazz bands, along the lines of 'Saints'. John, at the time, was fighting to save his place in the Elysian Jazz Band. "It may encourage 'skiffle' groups," he wrote, "which after all enjoy themselves more than they give others pleasure, to enlarge their scope." This is ironic, as John, ejected from the Elysian, was shortly to become skiffler-in-chief at Unity Theatre and spread untold happiness.

Where were we?

An ad in *Sing* vol. 2 no. 4, (Oct.-Nov. 1955), shows that Topic's release schedule of Autumn 1955 is confined to 78s: they sell at six shillings apiece. Paul Robeson is well represented (TRC 94, TRC 95, TRC 96), whilst Pete Seeger and Ewan MacColl have one apiece (TRC 92 and TRC 93 respectively). The prefix TRC is a retention from the Topic Record Club, which started sending a 78rpm disc a month to every WMA member back in 1939, before it got derailed by war.

The shock of the new is heralded by an ad in *Sing* vol. 2 no. 6 (Feb.-Mar. 1956), when Topic announced a series of 8" 33⅓ microgroove long-players, and apologised for the delay of T1, T2, T3 and T4. Some of the titles are unfamiliar. T6 is listed as *Off to Sea Again* by Ewan MacColl and A.L. Lloyd; presumably the material ended up as T7 and T8, *Row, Bullies, Row* and *The Blackball Line. Songs of Woody Guthrie* is simply the working title for *Woody Guthrie's Blues* by Jack Elliott, with the same catalogue number, T5. *Irish Songs and Dances*, originally designated T7 and trailed as "old and famous ballads, and some newer ones, and traditional fiddle and pipe tunes to set you dancing", had to wait until 1959 to see the light of day, and then as 10T6 – *Street Songs and Fiddle Tunes of Ireland* by Margaret Barry and Michael Gorman; Willie Clancy, co-billed in the early ad, had fallen by the wayside. Something called *In the Fight For Spain* and assigned T8 failed to appear altogether. Did Topic lose heart at another lost cause? At any rate, the twentieth anniversary of the fight against fascism in Spain passed unmarked by Topic. Amongst these rumoured records and unrealised projects, no mention is made of the batch of 12" 33⅓ long-players that materially existed and could, theoretically, be purchased. Or, as Bill says, it was more, "you had to know someone who already had one, and try to buy their copy,"[4] such was the arbitrariness of the distribution. I tell a lie: the unissued *Off to Sea Again* is mentioned as the follow-up to *The Singing Sailor* (TRL3). Otherwise TRL1, TRL2 and TRL3, the trio of 12" LPs that bridged the gap between 78s and the 8" long-playing oddities, are shrouded in mystery. TRL2 is more rumour than fact.

The review section of *Sing* vol. 3 no. 1 (Apr.-May 1956) defaults to 78s (TRC 97, TRC 98 and TRC 99), whilst T2, *International Folk Song Contest*, one of the 8" LPs delayed in February, gets a review in August (*Sing* vol. 3 no. 3, Aug.-Sept. 1956). The 8" long-players *Woody Guthrie's Blues* (T5), *Row Bullies Row* (T7), *The Blackball Line* (T8) and two volumes of *Irish Songs of Resistance* (T3 and T4) are promised before Christmas (*Sing* vol. 3 no. 4, Dec. 1956-Jan. 1957), which explains the expectant headline, 'Elliott for Christmas'. All the titles go on to receive reviews in *Sing* vol. 3 no. 6 (Feb.-Mar. 1957).[5]

A future format, the 7" extended play at 45 rpm, was introduced in 1959. Some early Topic EPs (TOP45, TOP46) were leased from French label Chant du Monde, early specialists in world music. Chant du Monde, informs Bill, put out dozens of 7" 33⅓ long-players, a common format on the continent. Whereas Topic's 8" 33⅓ long-players are unique.

The price differentiation in Figure 4 is reasonable enough: 12" discs require three times more plastic than 7" discs. Bill crunches the numbers. If, at 1957 prices, a Topic 78rpm disc retailed at 6s. 6d. and a

4 Bill Leader, annotation, Jul 3, 2020.

5 There is no *Sing* vol. 3 no. 5 due to Eric Winter's haphazard system of numeration.

Topic 7" long-player cost 16s. 8½d., "you could sell one hundred 6/6d records and make three quid or something. (Have I got that right? Six shillings … My arithmetic's not good.) But if you sold one hundred records at 16/8½d each you would make considerably more. I'll have to do the maths and sort that one out."[6]

Eric Hobsbawm wrote *The Jazz Scene*, a book about his favourite music, under the pseudonym Francis Newton. As befits the greatest Marxist historian of his age, he homes in on jazz economics. In one chapter he breaks down the cost of making a long player at current (1958) prices:

			s.	*d.*
Selling price (1958)		per copy	30	0
Cost per copy:	Distribution (52½%)		11	3
	Sleeve		2	6
	Pressing material, labels		2	0
	Artist's Royalty (5%)		1	1
	Mechanical Royalties (6¼%)		1	4
Musicians (10 men for 2 sessions), £104 at 1,000 sales				
		per copy	1	3
Studio costs (£80 at 1,000 sales)		per copy		11
Purchase Tax			7	6
Maker's margin, overheads, publicity			2	2
Break even point: say 1,200 sales				

Fig. 5 *The costs of making a jazz LP in 1958, courtesy of Eric Hobsbawm*[7]

Break even point: say 1,200 sales. This is a tall order.

"We couldn't have done it at those sorts of costs. God forgive us, we didn't pay MU rates, because we were dealing with individual artists, who were just paid a royalty on sales. If you didn't sell anything, you didn't pay them anything. We didn't have studio costs. Well, not normal studio costs. They were down to a minimum.

6 "LPs sold for £2 and 78s sold for about six shillings… Thus the maximum income possible from the release of a 78rpm record was £29.70, whereas the income from an LP was a stunning £198." Bill Leader, quote cit. Michael Brocken, *The English Folk Revival* (Ashgate, 2003), p. 59.

7 Francis Newton (Eric Hobsbawm), *The Jazz Scene* (Penguin edition, 1961), p. 173.

"And we did all sorts of things to try and keep the cost of sleeves down, with silkscreen, made-up sleeves, standard designs and what have you. The big companies were turning out full colour sleeves, and that's what people expected. We could come up with good designs, but we couldn't necessarily come up with the quality of production. We tried all sorts of ways of getting around that one. We did get around to paying graphic artists, eventually. But they were always sympathisers. They were always very generous and kind and easy on us. I mean, we didn't pay Gloria."[8]

There is no evidence to suggest that Hobsbawm strayed far from jazz in his listening habits, and why should he? Chronicling the twentieth century and searching for the perfect triple-beat are heroic pursuits, and not incompatible pursuits. In settling on 2,000 sales as average, he commented "this suggests that some LPs can still be made very cheaply", tacitly acknowledging that many jazz releases fell short of the target. For Topic, a sales figure of 2,000 was fantastical; far more realistic to settle for the Inland Revenue-prescribed 99 copies.

The small print-run, however, brought its own problems. Record pressing plants and distributors disdained a a micro-label like Topic because "our pressing requirements were too small. Our runs were too small to be economical for them."

And so Topic turned to Homophone, a plastics manufacturer with a plant in New Cross, South London. Its core business was the production of dolls and dominoes, and the quality of the raw material fell far below the demands of the fifties audiophile, much less his modern-day counterpart. Every small label had this problem. Argo used a firm in Mitcham called Tranco (the Transcription Company). Dick Swettenham, who worked for Argo at one point, told Bill a salutary tale about the 23rd recut of *Moonlight Sonata*, "played by a very distinguished pianist". It was "the 23rd attempt of the company to produce a record that didn't sound as if it was an old 78". Argo were eventually driven into the arms of Decca, which resulted in improved pressings at the cost of independence and control.

"Our pressings at Topic were pretty rough, but it was rough music in a way, so it didn't matter quite as much. We did survive with substandard pressings for a long time. I'm talking about '55-'60 now."[9]

---

8 Bill Leader, interview with the author, June 15, 2012.

9 Bill Leader, interview with the author, September 8, 2009.

A die is a machine-driven tool that shapes something by stamping it out of gross material. Dies at Homophone were designed for discs with very freakish diameters. For example, 2" discs, reserved for the insides of dolls, which, when tilted sideways, would exclaim 'Mama!' or some such. Crucially, the die used for pressing 8" shellac discs was still in working order. Historically, 8" discs were the preserve of Eclipse, Woolworth's house label. The extra width gave one more minute of playing time, giving Eclipse an advantage over its market rivals, which alike sold for sixpence. Its sister label, Rex Records, sold for the same price.

"Do you remember Rex Records?" asks Bill, mistaking me for an older man. "A black label with a cursive REX. Gracie Fields did some things for them. They sold in Woolworth's for sixpence."[10]

The substitution of vinyl for shellac was the only modification needed to get the Homophone die machine up and running. Thus, Topic ushered in the modern age by borrowing a pre-war format from 'Woolies'. The 8" diameter, says Bill, was "probably not a sensible idea, because the pick-up would come down automatically at a 12" diameter – or a 10" diameter, or a 7" diameter," but not an 8" diameter. Listeners were forced to put the disc on manually or forego the first track.

"The idea, the crazy idea behind it, was that we could get an 8" LP running at 33 rpm, not 45 rpm, so you could get on as much as you could on a 10" LP. A 10" LP was a valid format then. It died when the Americans discovered they could sell the same material for much more if they put it on a 12" record. People were sucker enough to buy it. So you hardly see any 10" LPs.

"But 8" LPs almost had the playing length of a 10" LP, and you could charge a reasonable amount on the basis of playing time. What you're paying the manufacturer for is the material you use, and you're actually using considerably less material for an 8" LP than for a 10" LP, or a 12" LP. The pressings are cheaper because they use less material. That was the theory behind our strange 8" LPs. The first lot we released were 12" LPs. Then we had a flood of 8" LPs, but then we settled down and became fairly normal with 10" LPs, 12" LPs and 7" EPs."

The odd freak still slipped through. *Round and Round with the Jeffersons*, by the Jeffersons (7T19), from 1958, had a 7" diameter and

10 Bill Leader, interview with the author, January 15, 2016.

played at 33⅓ with three tracks a side. Similarly, *Come Along John* by Peggy Seeger and her sisters (7T18) is attractive but confusing, being either a generous EP or a stingy LP, according to one's humour.

Swimming against the tide can be exhilarating, until you tire and drown.

"They were joyous days. We tried to cut our own LPs. We thought it would be cheaper than getting them cut at some studio. It cost a bloody fortune to get masters cut. So we bought this cutting machine that was made in Brighouse, by a fellow called Arnold Sugden. A genius. We were one of the few people in the world that bought his cutting machine. And then we spent the rest of the year trying to learn how to live with it, and coax stuff out of it. There were not many things that got issued that we actually cut. So that was an experiment that didn't work."

Bill, pressed for detail, explains: "There's so many things to go wrong. Sean Davies is probably now the only person in this country you could turn to, if you had a cutting machine and wanted to get it fixed. They were brilliant people, who understood all the problems and the solutions, and they worked on machines that cost thousands and thousands of pounds. Beautiful machines: well designed and efficient. Sugden's machine, of course, cost a couple of hundred quid – it was an amazing thing – but you had to do all the mechanised operations by hand."[11]

Sean Davies has sixty years' experience of disc cutting, and there is nothing he doesn't know about the machines and their maintenance. He cut discs at home as a schoolboy! (Sean's father worked for the company that acquired the MSS lathe company and took advantage of staff discount on lathes and blank lacquer discs and cutting styli to indulge his son's engineering obsession.) He joined IBC Studios as a junior engineer in 1958. Sean was in the same line of work as Bill and shared an interest in folk music. He is the engineer responsible for *Folksound of Britain*, CLP 1910, for HMV; later he passed his studio at Cecil Sharp House to Bill. This was where Sean recorded *Anne Briggs* (12T207, 1971), and Bill recorded *Bright Phoebus* (LES 2076, 1972). But I digress.

"Yes, Sugden had a tendency to make his machinery rather fragile," pronounces Davies with casual authority. "It was nicely made, and he

11 Bill Leader, interview with the author, June 12, 2015.

was a very clever bloke, but it was appropriate to what you might call a careful hi-fi environment. Now the studio world is not that. The studio world hammers machines, and they've got to work it day in, day out, all the time. It's very tight tolerance and it's no good having a machine which is off the air for two or three days every third week. You can't do business like that. There are people who say, right, I want that cut and I want that cut today, so you do it today. And that is the big difference between the hi-fi world and the studio world."[12]

The sleeves of 78rpm disc were not unadorned or uniform, as Bill attests.

"They weren't all plain by any means. Some were custom printed for the record shop that supplied the disc. Some were the product of the record label. I remember Walter Crane-ish, art nouveau, willowy women gracing many a record turned out by the Columbia Gramophone Company."[13]

The modern age, however, demanded long-players in cardboard sleeves of eye-catching design. Their manufacture, according to Bill, gave Topic it's biggest single financial headache. The company's ex-production manager gives a hard lesson in Topic economics:

"Records were reasonably economical. In fact, they were deliberately priced so that an enthusiast, or a small company, could afford to make one hundred records and just about recoup their costs. The manufacturers set their charges like that. But the initiation costs of printing sleeves meant that it was only worthwhile if you printed quite a quantity. We weren't producing quite a quantity of anything. We were living in the culture of one hundred and no Purchase Tax. It was still very much in our minds."[14]

Put simply, Purchase Tax was levied on the wholesale price of 'luxury' goods and was *initially* set at a rate of 33⅓ but fluctuated depending upon the degree of 'luxury' as defined by the Chancellor. Typically, records were loaded with 60% Purchase Tax, which, at odd times, rose to 70%. It fell upon the wholesaler to act as tax collector for the government. The system was superseded by VAT on the accession of the UK to the EEC in 1973.

---

12 Sean Davies, interview with the author, April 19, 2018.

13 Bill Leader, annotation, April 13, 2020.

14 Bill Leader, interview with the author, February 14, 2014, and for the rundown on the printing problem that follows.

"The tax wasn't charged on what you paid your supplier, which is how VAT works," explains Bill. "With Purchase Tax there was an army of Customs and Excise officials, who, if you were registered for Purchase Tax, would come and take over your office or your accommodation for a day or two, and go through all your books and then tell you how much Purchase Tax you had to pay. If you weren't registered for Purchase Tax – and you didn't need to be if your turnover was modest – then you had to pay Purchase Tax based on what the final selling price would be.

"So you were buying in a stock of, say, 1,000 sleeves at the manufacturers' price but adding the government's full whack. And if you didn't intend to use these for the next century or two, because you were producing a hundred records every now and then, it was very uneconomic. Which is what we were trying to cut down on. We didn't want to spend all that money on something that was not important. Not important, because we weren't selling to anybody who was going to buy the record on the basis of 'Wow! What a sleeve!' We weren't Blue Note. The people who were going to buy our records were going to buy our records, and the rest of the community were not, irrespective of what sleeve we put on.

"So we tried to cut the cost of the sleeve. And it came down to buying ready-made white sleeves and getting my friend, the Ilford stationery rep, to rig up his clothes-line and burn the midnight oil with his silkscreen."

At a further session, Bill divulges the name of the Ilford stationery rep, and talks about the circumstances of their first meeting.

"I used to work late at Bishops Bridge Road. Sometimes it wasn't as late as others. One night I thought, 'I'll just go and see what's on at the Metropolitan.' Because sometimes you could sneak in and get a seat once the show had started, and if it was a disappointment, you hadn't wasted any money on it."

The Metropolitan, a music hall on Edgware Road, was an easy walk from the WMA offices in Paddington and, just as Bill had anticipated, the box office was shut for the night.

"The first half was going on. I went through the doors into the stalls, and I was just looking around to see if there was a seat when this uniformed person came across and said, 'Can I help you?', which is the polite way of saying 'What the bloody hell are you up to?' He said,

'Look, there's a bar downstairs, you go down there. We've got a few more minutes to run before the interval, and when everyone comes up from the bar, you can join them.' He also turned up at the interval, and I obviously had to offer him refreshment. That's how I found John Alexander – Alex, we called him."[15]

(On a technical note, my method of organising material is to meticulously transcribe interviews and then arrange according to theme, or a thread of logic that suits my purpose. Naturally, there's a degree of artifice involved, and I'm risking something discussing it now. The idea is to maintain the illusion of a single interview unfolding in the present moment, and the matter under discussion is in the past tense. In at least one place I've spliced together quotes from separate interviews in a single paragraph. For reasons of documentary integrity, footnotes give the source of the quotes, and the dates of the interviews. In this way, my aim is for past and present to connect in a linear way, which is what past and present are supposed to do. Grammarians call it 'the present perfect'. It suits the film medium well: we accept cuts between speakers in a documentary without question. It's trickier in the written medium. Organising Bill's comments in a seamless flow is straightforward enough, but when a range of voices enter, and coherence must still be maintained, well, strains can show. If the holders of some of the voices have died in the meantime, I can feel like a medium in a seance.

This is very strange, not to say fearful, but I'm committed to the present perfect for sentimental reasons. There will come a time, not far off, when Bill will no longer be with us – it is his 91st birthday as I write – and then, at a time more distant but still a blink of the eye in the grand scheme, I'll be gone too. It's consoling to know that somewhere Bill will still be patiently explaining to me the difference between a jig and a reel. The difference, by the way, is a matter of time. Joyce's exegesis on the last page of *Dubliners* expresses what I'm talking about, *temporality*, in a profound way: "His soul had approached that region where dwell the vast hosts of the dead…"[16]

Bill thinks that future generations will bless my name. The faint sardonic trace, conveyed with brimming geniality, is a characteristic Bill trait. The inference being, *for all the good future generations will do you!* Me, I'm just grateful for an encouraging word.

15 Bill Leader, interview with the author, November 28, 2014.

16 James Joyce, *Dubliners* (1914; Penguin Modern Classics edition, 1962), p. 220.

I mention this now because a dead man is about to enter.)

"I don't remember that," says John Alexander, about the encounter at The Metropolitan, honouring the convention that no-one remembers meeting Bill for the first time. "But if he says so, that's probably true. A lot of people tried to sneak in because they knew me, and I didn't always let them in."[17]

John Alexander's Metropolitan memoir, *Tearing Tickets Twice Nightly*, mentions how John Foreman, the junior doorman of the Metropolitan, perpetrated a scam involving tearing a ticket in two and giving half each to paying customers, in this way reserving complete tickets for his mates from Unity Theatre to get in for free.[18] If Alex was complicit in this fraud he isn't telling.

"John Foreman wasn't working there to begin with," he says, avoiding the question. "I started and about a year afterwards John Foreman came along."

Alex was also involved in Unity Theatre.

"Oh yes, very much involved. My wife at that time was the secretary of Unity." (Joyce Alexander; I might have something to say about strong communist women later – Bill married two of them.)[19] Alex himself was chairman of the music hall at Unity.

"Unity were well-known for their music hall. If ever there was a delay in putting a production on they used to shove music hall on as a stopgap," Alex explains.

His story is that he was made chairman as the only person with a gavel.[20] John Foreman has a less disingenuous explanation.

"Alex didn't think he could sing, but he liked being in charge, so he liked being chairman. He was a pretty good chairman of Unity music hall. Unity was marvellous music hall."[21]

---

17 John Alexander, interview with the author, December 9, 2015.

18 *Tearing Tickets Twice Nightly: The Last Days of Variety* by John Alexander (Arcady, 2002), pp. 21-22.

19 Bill's comment upon reading: "I decided not to plead for my two communist wives to be clarified as consecutive rather than coincidental, assuming it will be taken as red."

20 'In Memoriam – John Alexander', *Slapstick: Celebrating the History and Entertainment of Punch and Judy*, no.3, December 2019. https://issuu.com/punchandjudy/docs/slapstick_3_final_2019_online_edition.

21 John Foreman, interview with the author, April 23, 2020, and for the following Foreman quotes.

By tradition, the chairman always obliged with a routine of his own, and Alex based his act on something he'd picked up in *Bunkum Entertainments*, a Victorian book of spoofs and games. He and Peter Jones concocted a mind-reading act so bad that it was good. "What you're enjoying is knowing how he's trying to cod you," says Foreman, nailing the appeal. "You know it's not really mind-reading, but it's so obviously not that it's funny."

This Peter Jones was a journalist on the South London press as opposed to the actor and wit mentioned in the last chapter, and the scion of a music hall family far-famed as 'The Flying Zadoras', a trapeze act where a girl would be fired as an arrow from a bow by a big elastic band and caught by a fellow acrobat. The name 'Zadora' was in his endowment, and thus the spoof mind-reading act became 'The Great Zadora'. Was John involved in Zadora?

"I think I was the first."

"Oh!"

"Peter Jones probably did more. Anybody could do it. It's just a matter of saying yes and no and doing stupid things."

I would like to introduce you to a great mind...

"I can't give you all the spiel, but the text is probably down there somewhere. All kind of comic things were going on. He'd blindfold the great mind. You would be playing Zadora. You'd sit down and he'd put a blindfold on and then he'd ask you questions. He would put his hand on your head...

'What am I touching?'

'A head.'

'Whose is it?'

'It's mine!'

"It was all a bit hilarious. One of the first things he would say was, 'first of all I'll attach an earthing chain to facilitate the source of electrical brain power', and he'd clip a lavatory chain to your coat with a big clank, to know that you were earthed. It was quite a good act, the Zadora act, yeah. Alex would present it." He was, says John, "lifelong pro-circus. And the last thing he was very interested in was lantern slides. He knew all about magic lanterns. He really took it seriously. He was an amazing chap actually."

"He had all all these splendid enthusiasms," agrees Bill. "He was a devout believer in the future of the airship. 'If you're going to move freight

Two Punch and Judy men – Prof. John Alexander and Prof. Joe Beeby at the Cutty Sark in Greenwich, 1971

around the world reasonably quickly, bulky stuff, the easiest thing is to hang it from a big balloon full of helium and fly it where you want to.' He's probably right. It may well happen. 'There are two sorts of airship,' he said. 'There is a) rigid and b) limp. Hence the word, 'blimp'..."[22]

The major enthusiasm, however, was Punch and Judy. If the formative event in young Elliot Adnopoz's life was a rodeo at Madison Square Garden,[23] with Alex it happened to be a Punch and Judy Show on the sands at Clacton-on-Sea (coincidentally, the two entertainments took place in the same year, 1940). Aged 11, John Alexander decided to be a Punch and Judy Man. He had achieved this by 1945 ("he didn't muck about," says Foreman, admiringly). As part of the Holidays at Home initiative – designed to discourage the public from travelling in wartime – Alex assisted a local Punch man called Will Hull, who taught him the rudiments of the business.

Murder is committed daily on Britain's beaches during the holiday season ('Seafront Horror as Man Beats Wife, Child and Policeman' in tabloid language) and in front of children too. Indeed, the spectacle is laid on specially for children. I have not been in the audience of a Punch and Judy Show since I was a child but strolling past model

22 Bill Leader, interview with the author, November 28, 2014.

23 Ramblin' Jack Elliott: his story is coming up in the next book.

theatres on sands or in parks, I have seen Mr Punch's impact on junior spectators. Uxoricide, infanticide, sausage-fancying, execution and damnation make for simple yet engrossing entertainment. The engagement is total: rapture mixes with anxiety at every squeak from Mr Punch. Perhaps the existence of evil, the old primordial mystery, will be explained if the kids are attentive enough.

The young Alex may have consciously moulded his character according to the job specification of the Punch man – being playful, fearless and a natural rebel. His wife Joyce held stronger convictions and was more politically committed and practical (she led a nursery campaign after the war, in the face of male, left-wing trade unionists who didn't see nursery provision as a priority). Alex was more a don't-let-the-bastards-grind-you-down type. One of the last of the unlicensed street performers, he pitched on Hampstead Heath near the donkeys and at Greenwich near the Cutty Sark. On Royal Avenue, Chelsea, he shared an illegal pitch with Punch man Joe Beeby. The proscenium arch of his model theatre was decorated with the Latin inscription *Honi Soit qui Garde ses Pense* – 'shame on him who doesn't pay.' He had cause to remember the words in his dealings with Topic's General Secretary Gerry Sharp.

Printing was a spin-off from his day-job selling stationery, because, says his son Max, "if you wanted a book of memos or you wanted anything specific for a firm, you had to go through someone who knew how to set it up for letterpress." When the British Peace Committee – a spin-off from the CPGB – needed graphics for some publicity, Alex diversified into silkscreen printing. "He ducked and dived in those days," Max adds.[24]

"There was a crisis at Topic because they couldn't afford the sleeves," explains Alex. "And I used to drink with Bill, and he told Gerry Sharp that I might possibly be useful in this direction because I'd learned to silkscreen print. And that's how I started.

"I did some posters in blue, fairly small posters, which was my first attempt. To tell you the truth, the ink was too loose, and they weren't very good. I was amazed and delighted to see them coming into Trafalgar Square at a big rally, held aloft by one of the parties from somewhere. I think they must have got them from my wastepaper basket."

Alex confirms the modus operandi of the Topic job.

"In my shed, yes. Six by four, it was. Cost me £120."

---

24 Max Alexander, interview with the author, April 20, 2020.

John Alexander on holiday
(collection: Max Alexander)

"It must have been the smallest silkscreen printing works in the country," says Max, who was Alex's assistant for extra pocket-money.

"We had clothes pegs in the roof," Alex explains. "They were nailed into the rafters of the shed..." (no clothes-line; Bill has that wrong) "... and they would hold enough sleeves to dry half the run. So we would fill the racks up, climb out the door underneath the sleeves hanging up, have a cup of tea, and by that time they'd dried, and we'd go back and take them off and do another lot."

And what would the print run be? One hundred, perhaps?

"No. Three hundred very often. Gerry Sharp was a very hard taskmaster. I remember on one occasion he put too many records and sleeves in a box and put the boxes one on top of the other. The result being that the sleeves stuck together, because of the ink... It was a gloss ink, and gloss ink never dries properly. And he insisted that I did the whole thing again for free."

(Max mentions that Alex would keep the screens, and the sleeves would be reprinted as the order came in. The runs were small.)

"Were you paid for your services or was it voluntary?"

"No, I was paid. Very badly."[25]

"He was supplying more than Topic out of that shed," says Max, and reveals how Alex had a sideline printing tricks for Davenports

25 John Alexander, interview with the author, December 9, 2015.

Magic Shop (est. 1898), including outsize playing cards and images in cardboard of rabbits and other objects that came out of a magician's hat.

"He was ducking and diving doing lots of things," says Max for a second time.

("The first thing I ever did with him was to go to a meeting of magicians, a Magic Circle crowd, somewhere in East London," remembers Foreman. "Is it called Seven Kings? He was terribly well-known in the magic world, and he knew several important magicians, American ones, Scottish ones, and the Davenports. He did a lot of magic.")

And, because silkscreens do not support text, Alex bought some litho press equipment upon Gerry Sharp's insistence, specifically to print the catalogue number on the back, usually in the top right-hand corner. Otherwise, the backs of the early silkscreens were left blank, or, if there were words, they were big and bold. The sleeve-notes would then be duplicated (very often by John Foreman on Unity's Gestetner Duplicator) and bundled inside the record bag along with the record. Sleeve-notes were duplicated, never printed.

> Looking at the *Barrack Room Ballads* sleeve [second edition, with the boot and rifle-butt; Alex's own design] you can see his skills had developed to be able to produce paper cut stencils for two-colour printing and he was able to use blocking medium to produce acceptable lettering. The Topic logo would have used an early version of Letterset which was how graphic designers produced professional lettering for print before computers. The point was that it was all very cheaply done.[26]

"It was very much a desperate attempt to keep records coming out when you had no bloody money, and the thing that used up the money was sleeves. Not records. We recorded it all for no more additional costs than my wages, because we borrowed or bummed the equipment off somebody," says Bill.[27]

In sum, Topic kept afloat long enough to become the world's oldest independent label by practicing frugality and relying on comradely, unhistoric acts. In particular, they a) circumvented the distribution network with a mail-order system, i.e., Topic Record Club, b) embraced novel formats such as the 8" disc to save on materials and manufacture

26 Max Alexander, email to the author, April 24, 2020.

27 Bill Leader, interview with the author, February 14, 2014.

costs, eventually coming around to the industry standard, and c) devolved jobs to enthusiasts in return for something Bill blithely describes as "the satisfaction of a good job well done" (and he should know).

As we have seen, sleeves constituted the biggest expense, and a way out of this trap was proposed by Jim Boswell, the artist and graphic designer, in the form of the stripey readymade. These were uniform jackets, all with a striped design, some on orange card, printed in bulk in both 12" and 10" formats and seem to have been reserved for reissues. It worked like this: each album was housed in an identical striped jacket, which, being uniform, could be inexpensively printed in bulk. Records were differentiated by a bit of paper pasted onto the sleeve, announcing title and catalogue number together with a quick but atmospheric sketch from Boswell, pertaining to the theme. *Shuttle and Cage*, which retained Ern Brooks' oily rag man, was the sole exception to the rule.

A second piece of paper with song titles and notes was stuck into the centre gatefold. Each stripey was a piece of card folded in the middle to make a gatefold – the simplest way to make a record sleeve. The inside was left blank. Orange was the favourite colour: white was glaring and monotonous. The bag for the disc was pasted onto the righthand side of the inner gatefold, and, because its open edge ran parallel to the middle fold, records were loaded from the left.

In a further refinement, only one long strip of paper was used, containing cover sketch and, printed upside down, the descriptive notes. The artwork was pasted on the front, and the notes hung loosely in the middle, right way around when folded. This strategy halved the paper and glue costs at a stroke.

Like Alex, James Boswell received a nominal sum. His daughter, Sal Shuel explains:

"He was a director of Topic. He was a cheap director, but he did get paid. He couldn't afford to do everything for nothing. He had to earn a living." (The payroll is growing; we started out with only two paid employees.) "The thing about Topic was, he lived at the time in Parliament Hill, and Topic was just around the corner in Nassington Road. They were within shouting distance of each other. If they wanted something they just came around and said, we need this, and he did it. He was very quick. He worked very fast."[28]

28 Sally Shuel, interview with the author, October 17, 2017.

The detail about Nassington Road gives a clue to the date of the stripeys. Topic Records moved from Bishops Bridge Road to Nassington Road in 1960. Another clue is the label. All the discs in striped sleeves have labels with the Fred Shapurji Topic logo (lower case 'topic' set in a black block, against a cream-coloured background). The famous blue Topic label was introduced in 1964. Thus, we can date the stripeys between 1960-64, which makes it doubly strange that Bill has no recollection of the stripeys. They're his records for the most part, and the period coincides with his time in record retail at a specialist folk shop. Sally Shuel remembers the stripeys well.

"This was Jim being efficient and business-like and professional. He was a graphic designer, as well as being an illustrator, and what he did was to create, very early on, an instantly recognisable style. It was a quick, easy way of doing it. And if he wasn't around and doing something else, other people could do it using his basic background. He did that quite often. At one point he completely redesigned *The Economist*.[29] What he did was give them a basic set-up – type-sizes; how wide the column should be – and said, just get on with this. It's only mathematics after that.

"He did that with Topic. He gave them a way of doing it. If he wasn't there, which in actual fact he nearly always was, they just simply got on with it and did it."

But even Sally is hazy about the precise date. Alex isn't aware of striped sleeves so they can't be his handiwork, which was my first guess. The year on the label is misleading because it gives the year 'first published' which, nine times out of ten, is inaccurate. Why this should be so has caused much head-scratching between Bill and myself.

"Sometimes we took an awful long time to bring stuff out. That could explain the discrepancy," is the best Bill can offer.[30]

---

29 *The Economist* magazine has been in weekly publication since 1843, and it holds the political and business establishment in thrall to its god-like omniscience and absence of self-doubt. My media is more *London Review of Books*, where I read about *The Economist*: 'In Real Sound Stupidity the English are Unrivalled', *LRB*, vol. 42, no.3, Feb. 6, 2020.

30 Bill Leader, interview with the author, June 15, 2012. Those stripeys in full: 10T6, Margaret Barry and Michael Gorman, *Street Songs & Fiddle Tunes*; 10T14, Jack Elliott and Derroll Adams, *The Rambling Boys*; 10T15, Jack Elliott, *Takes the Floor*; 10T25, Ewan MacColl with Peggy Seeger, *Second Shift*; 12T41, Ewan MacColl and Dominic Behan, *Streets of Song*; 10T50, Ewan MacColl and Isla Cameron, *Still I Love Him*; 12T51, A.L. Lloyd, *Outback Ballads*.

The stripeys conceal their secrets. Like the sun and stars, they are just there. In fact, we are nearer to penetrating the mysteries of Mars than the mysteries of the stripeys.

It is the summer of 1956, as near as we can make (Bill never kept a diary). The future is still a blank card awaiting a silkscreen. Bill is doing some chores at the WMA office on Bishops Bridge Road when Ewan MacColl walks through the door. MacColl, a man of destiny, is free of modesty about his accomplishments. *He has a high profile.* He wants to talk to Will Sahnow urgently about something. Camilla Betbeder introduces him to Bill.

"This is Bill Leader who's just come down from Bradford…"

The meeting was deceptively insignificant. No-one remembers first meeting Bill, but Ewan MacColl willed himself to forget it the moment it happened…

APPENDIX

Possibly Significant People

Stanley Accrington (picture courtesy of Stanley Accrington)

ACCRINGTON, STANLEY
Born ("as somebody else"): 1951, Coventry

Stanley is a compulsive song-writer who has been at it since the late 1970's when an Iron Lady annoyed him intensely; moved to the North of Greater Manchuria in 1978 when he became the last Station Master based at Rochdale railway station (and would like to apologise for any inconvenience to your journeys); his railway travels gave him great scope to indulge in his other profession, anagram maker; he has sung his weird and strange and sometimes serious songs in places as far as Hong Kong, North Carolina, Guernsey and Cleckheaton, and many places in England that most people wouldn't get to; he often heard of a legendary figure called Bill Leader on his musical journeys, but never imagined he would meet him at his favourite pub in Middleton; Stanley has been asked to open a moorland restaurant's septic tank, has appeared on a *Desert Island Discs*-type programme in Hong Kong the week following Kylie Minogue, and has been the answer to a quiz question in the Rochdale and District Quiz League; he is now living in Taxexile, an Aztec settlement on the edge of the Pennines, where he still writes songs, and where he grows raspberries, rhubarb and old.

(Biography kindly supplied by Stanley Accrington; note the textbook example of a zeugma in the last line.)

Chris Ackroyd (picture by Eva Navarro)

ACKROYD, CHRIS

Born: Mar 2, 1949, Keighley, Yorkshire

Gardener, *vendangeur*, artist's model, art historian, a friend of the author

Leaves school at 16 and has fifty jobs before age 30 inc. 'parks gardener and gravedigger', nanny, night-shift worker at Rustless Iron Co. (TRICO), where he climbs the chimney and has great difficulty getting down; the casual jobs earn money to travel; with his great friend, Eric Leech a.k.a. Youngfellow, he lights out for Europe, and follows the season as a *vendangeur* (grape picker), 1971-76; a youth club worker when punk breaks; an extra in *Yanks* (1979); studies History of Art as a mature student at Manchester Polytechnic, 1979-82; earns an MA in the same, 1982-83; lectures in History of Art at Manchester Polytechnic, beginning January 1984; an extra in *Scandal* (1989) and lead in *The Glove* by Ulric Finje; poses as Christ in artist Ghislaine Howard's *Stations of the Cross*; explores the impact of the Italian Renaissance on nineteenth century British art in his MPhil thesis, *An Examination of Writings on Michelangelo, 1750-1920*, which the current author types up, *c.* 1990; retires to West Yorkshire, 2014.

Book: *The History and Techniques of the Great Masters: Toulouse-Lautrec* (Quarto, 1989)

ALEXANDER, JOHN
Born: Mar 6, 1928, London
Died: Sept 28, 2019, Southend
Punch and Judy man, doorman of the Metropolitan music hall, stationery rep and silkscreen printer

John – familiarly called 'Alex' – learns the art of Punch from Will Hull and is performing his own show by the end of the Second World War; he fly-pitches on Hampstead Heath near the donkeys, and Greenwich near the Cutty Sark, and alternates with Joe Beeby at the Royal Avenue in Chelsea; John Foreman is his bottler before son Max takes over; chairman of music hall at Unity; creates 'The Great Zadora', a cod mind-reading act with Peter Jones; a stationery rep and printer by day, he silkscreens Topic sleeves by night, assisted by Max; meets Charlie Chaplin in his capacity as doorman at the Metropolitan; establishes a printing business, Arcady Press; co-founds the Punch and Judy Fellowship with son Max, 1981; moves to Southend and performs for several seasons on the beach at Broadstairs; self-publishes original and reprinted books under the Arcady imprint, inc. *Tearing Tickets Twice Nightly*, a personal history of the Metropolitan, 2002, and *The Expanded Frame File*, 2003.

Personal: m. Joyce (1949-separated ?); three children, Max, Jimmy and Clare; m. Julie Avery, 2002.

Quote: "Good god! I really enjoyed that!" – Arnold Wesker, overheard talking to his three children after a performance of Punch and Judy by 'Professor Alexander'.

Dave Arthur (picture courtesy of Dave Arthur)

ARTHUR, DAVE
Born: Nov 20, 1942, Little Sutton, Cheshire
Singer, musician, writer, researcher, presenter, Fellow of the Royal Society of Arts

Leaves home to avoid taking O Levels at St. Olave's Grammar School in Orpington and moves to Soho where he makes a bare living busking (w. Clive Palmer a.o.), works in cafes and restaurants and book shops (inc. at Collet's 'Bomb Shop'); meets Toni and moves to Oxford to run a university bookshop owned by Robert Maxwell, before inevitably falling out with Maxwell; resisting attempts to turn them into Sonny and Cher, Dave and Toni release four albums, *Morning Stands on Tiptoe*, Transatlantic, 1967, *The Lark in the Morning*, Topic, 1969, *Hearken to the Witches Rune*, Trailer, 1970, *Sing a Story*, Decca, 1977; edits *English Dance and Song*, the house magazine of the EFDSS, 1978-2000; a prolific freelance writer, researcher and broadcaster, theatre performer and storyteller (some of his ventures are listed below); with Pete Cooper and Dan Stewart forms banjo-driven old-time string trio Rattle on the Stovepipe, 2006; the author of *Bert*, the acclaimed biography of A.L. Lloyd, 2012; arranges Shirley Collins' comeback album, *Lodestar*, 2016.

Selected TV scripts/BBC Radio drama documentaries as writer/researcher/presenter/musician: *Playschool*, *Playaway*, *The Romans in Britain* (1983), *Middle English* (Thames, 1983-6), *Seeing and Doing* (Thames, 1985-92), *The DJ Kat Show* (BSKYB, 1988-95), *Patchstop* (1995).

Playwright (with Toni Arthur and Dave Wood), *Robin Hood*, Nottingham Playhouse, Young Vic, 1983; *Jack the Lad*, Manchester Library Theatre, 1984, *The Pied Piper*, Orchard Theatre, Plymouth, 1989; playwright/ writer in residence at Cannon Hill Puppet Theatre.

Journalism: *English Dance and Song Magazine* (editor, 1984-2003); *Society of Fight Directors Magazine* (ass. editor, 1985-6), *Animations* (National Puppetry magazine, editor, *c.* '90s), *Storylines* (magazine for the Society for Storytelling, editor, *c.* '90s); contributor to *The New Grove Dictionary of Music and Musicians*, *The Book of Music*, *Caxton Encyclopedia*, *Melody Maker*, *The Stage*, *Arts Desk*, *Words International*, *Folk Music Journal*, *Encyclopaedia Britannica*, and various poetry publications; reviews and writes obits for *The Guardian*, *The Independent*, *The Times*, and many regional newspapers and magazines.

Personal: m. Toni (1963-1977, separated); two children, Jonathan and Tim.

Awards: The EFDSS Gold Badge, 2003.

Books: *A Sussex Life - Gilbert Sargent, Countryman* (Barrie and Jenkins, 1990); *Bert - The Life and Times of A.L. Lloyd* (Pluto Press, 2012); *Memento Mori* (poetry, Chapel Press. 2020).

ASCH, MOSES ('MOE')

Born: Dec 2, 1905, Warsaw, Poland
Died: Oct 19, 1986, New York City, USA
Engineer and Folkways Records chief

US recording engineer and record executive of Jewish Polish heritage; one of his early jobs is installing sound systems in Yiddish theatres on the Lower East Side, NY; the first disc on Asch Records is his father, writer Sholem Asch, reading Bible stories; Asch Records becomes Folkways Records in the LP era; records NY-based talents Woody Guthrie, Leadbelly, Pete Seeger; documents dissent from the Civil Rights struggle (*The Story of Greenwood, Mississippi*, FD 5593) to the Scottish anti-Polaris campaign (*Ding Dong Dollar*, FD 5444) and develops a niche in the academic market: ethnographers send their field recordings for release in Folkways' distinctive heavy cardboard sleeves, with a divider to accommodate a booklet; the remuneration is small but the experience is mutually satisfying; the ethnographer gets a line in his CV and Asch has an addition to the catalogue; in this wasFolkways acquires a reputation for academic rigour and abstruse esoterica; the label's catalogue of some 2,100 recordings is acquired by the Smithsonian Institute after Asch's death, and honours his wish that no record be deleted.

The wisdom of Moses Asch #1, his Declaration of Purpose: "My obligation is to see that Folkways remains a depository of the sounds and music of the world and that these remain available to all. The obligation of the company is to maintain the office, the warehouse, the billing and collection of funds, to pay the rent and telephone, etc. Folkways succeeds when it becomes the invisible conduit from the world to the ears of human beings."

The wisdom of Moses Asch #2: "Just because the letter J is less popular than the letter S, you don't take it out of the dictionary.

James Boswell (picture by Brian Shuel)

BOSWELL, JAMES ('JIM')
Born: Jun 9, 1906, Westport, South Island, New Zealand
Died: Apr 15, 1971, London
Painter, illustrator, activist

His father comes from Scotland, a schoolmaster, and his mother from New Zealand; moves to London to study at the Royal College of Art, 1925-29; co-founds the anti-Fascist pressure group, Artists' International Association, or AIA; organises exhibitions in support of the Republican cause in the Spanish Civil War; contributes cartoons to the *Daily Worker* as 'Buchan'; becomes arts editor of the *Left Review*; joins Shell Petroleum Company, 1936; serves in the Royal Army Medical Corp during the war and is stationed in Iraq, 1942-43; debarred from being an official war artist because of his radical politics, the War Artists' Advisory Committee nevertheless collects his work; posted to the Army Bureau of Current Affairs, or ABCA, the education wing, 1944-45: ABCA has been credited, in part, for Labour's landslide victory in 1945; returns to Shell upon demob, 1944; resigns from Shell ("now more clearly than ever a powerful buffer for state monopoly capitalism" – Graham Stevenson), 1947; art editor of *Lilliput* magazine, 1947-50; designs film posters for Ealing Studios, 1947-54; mural painter for the Festival of Britain, 1951; edits and illustrates *JS Journal*, the staff magazine of Sainsbury's, 1951-71; director and designer for Topic Records, 1960-71; the chief art designer of the Labour general election campaign ('Let's Go with Labour'), 1964.

Personal: m. Elizabeth (Betty) Sears (1933-66, separated); daughter Sally, b. 1936; partner of Ruth Abel (TV producer, writer and publisher) who changes her name to Boswell by deed poll in 1967.

Selected exhibitions: a centenary exhibition, Tate Britain, 2006; an exhibition of war drawings, British Museum, 2006.

Book: *The Artist's Dilemma* (New Developments #1, 1947).

Further reading: William Feaver, *Boswell's London: Drawings by James Boswell Showing Changing London from the Thirties to the Fifties* (Wildwood House, 1978); William Feaver, *James Boswell: Unofficial War Artist* (Muswell Press, 2007).

BRIDGES, HARRY
Born: Jul 28, 1901, Victoria, Australia
Died: Mar 30, 1990, San Francisco, USA
Union leader, president of the International Longshoremen's and Warehousemen's Union (ILWU), 1937-1977.

Becomes a maritime seaman at the age of 16; settles in San Francisco as a longshoreman, 1922; a member of the International Longshoremen's Association (ILA) and leads an ILA dockland strike in San Francisco that escalates into a city-wide general strike as a response to 'Bloody Thursday', where police violence leaves two dead; survives assassination attempts and efforts to have him deported; elected president of a new union, International Longshoremen's and Warehousemen's Union, which affiliates to the Congress of Industrial Organizations (CIO), 1937; the Almanac Singers lionise Bridges on 'Ballad of Harry Bridges', 1941.

Quote – "The trouble Harry Bridges had on the West Coast took place while I was making various noises on the radio there in Los Angeles and, well, I just sort of thought they ought to be some kind of a little song wrote up about old Harry and the tough old human race for which he stands" – Woody Guthrie.

BRINSON, PETER
Born: Mar 6, 1920, Llandudno
Died: April 7, 1995
Dance writer, lecturer, educationalist and filmmaker

A man of dance, Brinson's achievement is to raise the profile of dance in the UK; his Oxford education is interrupted by the Second World War; serves as a tank commander with Montgomery at the Battle of El Alamein, 1942; graduates from Oxford in 1948; writes and produces *The Black Swan*, the first ballet film to use stereoscope, 1952; with John Minchinton, manages Films of Poland, an initiative to propagate Polish film in the UK, based in the Polish Cultural Institute, Portland Place, c. mid-fifties; with John Minchinton, joint editor of *Films and Filming*, c. mid-fifties; Director of the Royal Academy of Dancing, 1968-69; creates 'Ballet for All', a touring group to promote ballet in the regions, 1964-79; Director of the Gulbenkian Foundation, 1972-82; the go-to man on art and public policy, circa 60s and 70s.

Books: *The Choreographic Art* (1963, w. Peggy Van Praagh), *Background to European Ballet* (1966).

BRYCE, OWEN
Born: Aug 8, 1920, London
Died: 2015, Northampton
Cornet-player and trumpeter, "day one pioneer of British jazz" (Pete Frame).

As a member of George Webb's Dixielanders, Bryce plays what is considered to be the first traditional jazz gig in the UK at The Red Barn in south east London, under the auspices of the Bexleyheath & District Rhythm Club,1943; his partners in the Dixielanders are pianist George Webb, clarinetist Wally Fawkes, trumpeter Reg Rigden, and, briefly, trumpeter Humphrey Lyttelton; opens The Hot Spot, a record shop in the basement of his radio shop in Woolwich, and consents to let regular customer Chris Barber play in the band – he draws the line at banjo player/singer Lonnie Donegan, whose voice he dislikes; jazz tutor at the WMA Summer School, mid-fifties through to the mid-seventies; shuts up shop and moves to a small holding in Wroth Heath, Kent, playing on Sunday lunchtimes at his local pub; runs a residential jazz course on Bix, the barge-home he shares with wife Iris.

Quote: "The Dixielanders were true dilettantes, overtly and aggressively intransigent in their policy, and they believed that learning to read music would rob them of the jazz spirit."

Alan Bush (right) and friend (picture courtesy of Bill Leader)

BUSH, ALAN
Born: Dec 22, 1900, Dulwich, South London
Died: Oct 31, 1995, Watford General Hospital, Hertfordshire
Composer, founder and President of the WMA

Studies composition at the Royal Academy of Music, early twenties; studies piano with Artur Schnabel, Berlin, 1929-31; founds the Workers' Music Association to bring positive social change through music, 1936; galvanises the workers' struggle with pageants, choruses, and bespoke operas and large-scale compositions; conducts *A Festival of Music for the People* with Paul Robeson among the soloists, Royal Albert Hall, 1939; banned by the BBC during the Second World War; deported from British Guiana while researching his opera, *The Sugar Reapers*, 1957; influenced by the Viennese School of atonal music, Bush simplifies his style post-war, in line with his conviction that music should be accessible to the people; the Alan Bush Music Trust is founded "to promote the education and appreciation by the public in and of music and, in particular, the works of the British composer Alan Bush," 1997.

Selected works: *Piano Sonata in B minor*, Op. 2 (1922); *Dialectic for String Quartet*, 1929; *A Dance Overture*, Op. 12 (1935), *Symphony No. 1 in C*, Op. 21 (1939-40); *Three Concert Studies for Piano Trio*, Op. 31, 1947; *The Lascaux Symphony*, Op. 98 (1983).

Personal: m. Nancy Head, sister of art songwriter Michael Head, 1931; she writes the text for many of his vocal works, and the libretti for operas inc. *The Press Gang or The Escap'd Apprentice* (1946), *Wat Tyler* (1948-50; winner of the 1951 Festival of Britain opera competition), *The Spell Unbound* (1953), *Men of Blackmoor* (1954-55), *The Ferryman's Daughter* (1961), *The Sugar Reapers* (1961-64), *Joe Hill: The Man Who Never Died* (1970).

Quote: "I believe that music should reflect the national musical and cultural traditions of the composer's country comprising both the folk music and the previously composed music of the country."

Mike Butler (picture by Eva Navarro)

BUTLER, MIKE
Born: Dec 22 1958, Middlesbrough
Artist, music journalist, the author

Switches his musical allegiance from Pink Floyd to Bobby Womack, *c.* 1975; flunks a History BA, Polytechnic of North London, 1977-78; drifts into record retail at Out on the Floor (still only a stall at Camden Town market) and Honest Jon's, 1980-1983; gains a first class BA in Art, Manchester Polytechnic, 1986; part of Manchester Artists Studio Association (MASA) and its exhibition space, Castlefield Gallery, 1986-1992; editor of the gallery's house magazine, *Granby Row Review*, 1987-1991; jazz editor at *City Life* what's-on magazine, where he interviews Sun Ra and (*twice!*) Nina Simone, 1989-2009; *Manchester Evening News* buys the title and inherits the jazz editor; compiles listings and interviews musicians for *Metro*, 1999-2009; discovers the missing verse of 'A Whiter Shade of Pale' for *Independent on Sunday*, 1994; promotes UK tours by Abner Burnett, 1998 and 2001, the latter with co-star Johnny Moynihan; co-promotes a joint UK tour by John Fahey and Abner Burnett, 1999 – a rock 'n' roll disaster story of such magnitude that it rates a feature in *Mojo*.

Personal: m. Eva María Navarro López, Aug 25, 2016

COLEMAN, WILLIAM CALDWELL
Born: Oct 17 1884, Louisville, Kentucky, USA
Died: Jan 12 1968, Baltimore, Maryland, USA
Federal judge

Studies law at Harvard University and Harvard Law School, 1905-1909; heads a private practice, Baltimore, Maryland, 1909-1927; US Army private, 1918; called to the US District Court for the District of Maryland, April, 1927; rules on prohibition cases etc., 1929-1933; refuses citizenship to German and Italian nationals, 1941; chief judge for the District of Maryland, 1948-1955; a Republican on the Baltimore County Council, 1957-1959.

Personal: m. Elizabeth Brooke; four children, William Caldwell Coleman Jnr (1918-1945), Rev. Robert Henry Coleman (1920-1957), Susan Coleman (1923-2006), Elizabeth C. Mooney (?-2003).

CORUM, ALF
Born: 1890
Died: Dec 26 1969
Composer, violinist, charter member of the WMA

A musician and conscientious objector during the 1914-18 war; seized from the orchestra pit of the Finsbury Park Empire to be delivered to prison, according to Graham Stevenson's excellent *Communist Biographies*; editor of *The Winchester Whisperer*, the prison magazine, which existed in a single copy and was written with a needle onto toilet paper and passed from one prisoner to another; as writing paper was forbidden, this is also how he wrote music; wife Evelyn, in a pierrot troupe when they meet, with a founder member of the Communist Party; accompanies silent film until he loses the job with the talkies; runs a record stall in Croydon Market until the Depression hits and Alf becomes a specialist in pest control, playing in theatre pit-bands at night; his own compositions tend to be serious, as opposed to the light stuff he plays for a living, and draw praise in the forties and fifties; establishes the Tooting Tenants' Defence League; a charter member of the WMA; active in the Co-op movement and the *Daily Worker* Choir; editor of *Music and Life*, the music magazine of the Party, for most of the fifties and sixties; a contributor to *Marxism Today*; his daughter

Wendy Corum gives an eye-witness account of the 1955 Youth Festival in Warsaw in *Sing*, and goes on to make one of Bill's earliest recordings, 'Villikins and His Dinah' b/w 'The Coachman', 1957.

Personal: m. Evelyn (1920); two children, Wendy (1923-2012) and Peter (1925-1963).

Quote: "A man who all his life most genuinely loved music. On occasion, when somebody has tried to tell me that orchestral musicians grow tired of playing, I have quoted him as an example of one who never did" –Ruth Gipps.

Gloria Dallas, formerly Leader (picture courtesy of Bill Leader)

DALLAS, GLORIA (formerly **LEADER**)
Born: Mar 29 1931, Bradford
Artist, teacher and activist

Née Whittington; Gloria joins the Young Communist League at the age of 17, which is how she meets future husband Bill; studies for her Art Teacher's Diploma at Leeds, 1952-55; gains her first post as an art teacher, Canning Town, 1955; designs record sleeves for Topic and programmes for Unity Theatre, branching into set design/costume design for Unity and interior design for the WMA – she decorates the music room at Bishops Bridge Road; a prophet of the Campaign for Real Ale with the self-published pamphlet *Make Your Own Beer*, 1960; confronts Tony Blair over the Iraq War on TV, 2003.

Personal: m. Bill Leader (1952-63, divorced); son Tom born, 1959; Frances, 1962 (died after a few days); m. Karl Dallas (1965-2016, his death); adopts Stephen and Molly w. Karl.

Selected LP designs (all for Topic, apart from *Larks*, which is Xtra): *Street Songs and Fiddle Tunes*, Margaret Barry and Michael Gorman; *Irish Songs*, Dominic Behan; *The Rambling Boys*, Derroll Adams and Jack Elliott; various: *Songs Against the Bomb*; various: *The Larks They Sang Melodious*; illustrates *Singers of an Empty Day* by Karl Dallas, 1971.

Karl Dallas (picture courtesy of the Karl Dallas estate)

DALLAS, KARL
Born: Jan 29 1931, Acton, West London
Died: Jun 21 2016, Bradford
Journalist, songwriter and activist

Named Karl Frederick after Marx and Engels, and known as Fred in his early career; as a songwriter, journalist and human being, he is passionate in his politics and without limits in his sensibility; edits *Folk Music*, 1963-76; writes about folk but also Stockhausen, Pink Floyd, Frank Zappa, and Van Der Graaf Generator during a long tenure at *Melody Maker*, 1957-1980, where he also falls in love with Joni Mitchell; converts to Christianity, 1984; travels to Iraq as a human shield following the US invasion, Feb 2003; *Into the War Zone*, a 'musical tragicomedy' inspired by his experience in Iraq is produced by the Writers Company in Bradford, 2005.

Songs: 'The Family of Man', 'Derek Bentley', 'That Greedy Landlord' a.o.

Books: *Swinging London*, pub. Stanmore, 1967; *Singers of an Empty Day*, pub. Kahn & Averill, 1971; *The Cruel Wars*, pub. Wolfe, 1972; *The Electric Muse*, with Dave Laing, Robin Denselow and Robert Shelton, pub. Methuen, 1975; *Pink Floyd: Bricks in the Wall*, pub. S.P.I. Books, 1994.

Quote: "Most of us are branch libraries, but one or two are the Library of Alexandria, whose loss represents a great extinction of knowledge and history. Karl Dallas was one such. It wasn't just folk music he knew about" – *London Review of Books*

DAVIES, SEAN
Born: Feb 16 1938, Llanishen, Wales
Mastering, cutting and recording engineer and studio designer

Begins cutting discs at home at home aged 13; sound engineer at IBC Studios, 1958-65 – barring an interval in the research department of a Sheffield steelworks making magnets for Goodmans and Wharfedale speakers, 1960-61; part of the team that develops the mellotron; creates a recording studio in Cecil Sharp House with Peter Kennedy, 1965; founds SW Davies Ltd, specialists in disc-cutting lathes and studio design, 1979; retires after 53 years in audio, 2013; a freelance audio consultant in studio design and disc cutting, 2013-present; currently writing what will almost certainly be the definitive history of sound recording.

Awards: Special Recognition Award at the MPG Awards, 2014; Sound Fellowship Hall of Fame, 2007; the Order of Lenin.

Selected recordings as engineer and disc cutter: Charles Mackerras / Pro Arte Orchestra, *Janáček*, Pye, 1959; various: *Folksound of Britain*, HMV (LP), 1965; *Folksound of Britain*, HMV (EP), 1965; George Belton, *All Jolly Fellows*, EFDSS, 1967; Jimmy McBeath, *Wild Rover No More*, Topic, 1967; Janet Baker, *French Songs*, L'Oiseau-Lyre, 1967; J.R.R. Tolkien, *Poems and Songs of Middle Earth*, Caedmon, 1967; various: *Owdham Edge*, Topic, 1970; A.L. Lloyd, Trevor Lucas & Martyn Wyndham-Read, *The Great Australian Legend*, Topic, 1971; Anne Briggs, *Anne Briggs*, Topic, 1971; King Crimson, *Red*, Island, 1974.

DAVIS, ANNE (NANCE)
Born: Jun 8 1892
Died: *c.* October 1975

Née Leader, the daughter of Thomas and Louisa Martha; as a child, she abandons her babysitting duties as the eldest to run off and dance to a barrel-organ; as a grown-up, she is an obsessive card player and a champion at pinochle; the first Leader to emigrate to the USA in 1923.

Personal: m. Harry Davis, 1918; children Ronald (1922-26), Harry (1920-2004).

Tom Driberg (picture by Brian Shuel)

DRIBERG, TOM
Born: May 22 1905, Crowborough, East Sussex
Died: Aug 12 1976, between Paddington Station and the Barbican in a taxi
Socialist, MP, journalist, "an intellectual, a drinking man, a gossip, a high churchman, a liturgist, a homosexual..." (*Times* obituary)

Joins the Communist Party aged 15; a gossip columnist on the *Daily Express*, 1928; expelled from the CPGB on the evidence of Anthony Blunt, 1941; Independent MP for Maldon, 1942; joins the Labour

Party, circa 1944; serves on the party's National Executive, 1949; Labour censure him for neglecting his duties to report from Korea, 1950; chairman of the Labour Party National Executive, 1957-58; loses Maldon, 1958, and wins Barking, 1959; implicated in the 'Peer and Gangster' libel action brought by Lord Boothby against the *Sunday Mirror* – he and Boothby attend the same orgies; visits Guy Burgess in Moscow, and, compromised by a honeytrap, is pressured into spying for the KGB, 1956.

Books: *Guy Burgess, A Portrait with Background*, pub. Weidenfeld and Nicolson, 1956; *Ruling Passions*, an autobiography, pub. Cape, 1977.

Further reading: *Tom Driberg: The Soul of Indiscretion* by Francis Wheen, pub. Chatto, 1990.

Quote: "Tom Driberg is the sort of person who gives sodomy a bad name" – Winston Churchill.

Alex and Louise Eaton (picture courtesy of Louise Eaton)

EATON, ALEX
Born: Oct 31 1923, Bradford, West Yorkshire
Died: Dec 7 2011, Honley, West Yorkshire
French teacher, organiser, activist, founder of the Topic Club

Contracts polio aged eighteen months; attends Carlton Grammar School, Bradford, where he later becomes French teacher; joins the YCL, 1948; a mentor for Bill Leader, with whom he starts the Bradford branch of the Workers' Music Association, 1954; meets future wife Louise whilst standing in for a colleague at a French night class, 1955; founder and director of the WMA Singers, Leeds; founder of the Topic Club, the oldest active folk club in Britain, Bradford, 1956; a contributor to the *Telegraph & Argus*.

Memoir: *Dot and Carry One* (Apple Books)

EATON, FELIX
Born: Sept 10 1957
Retail interior designer, director of shopfitters Replan, car designer

A successful retail interior designer, Felix transitions to automobile design with the Eadon Green Black Cuillin, a retro coupé styled on his kitchen table and launched at the Geneva Motor Show, 2017; he has designed and launched another two cars since then, and the last received a Mayfair showroom.

EATON, LOUISE
Born: Nov 13 1935, Bradford, West Yorkshire
Dancer, activist, founder of Day of Dance

Joins the Idle and Thackley Amateur Dramatics as a chorus-girl and dancer, 1955; whirlwind romance with Alex Eaton, circa April 1955; joins Persephone, a ladies clog-dancing team, 1989; founder of the Day of Dance, a fund-raiser for Yorkshire CND, which becomes an annual fixture, 1972 to the present; retires from dancing, and takes up tenor horn and ukulele.

Personal: m. Alex, 1955-2011 (his death); four children, Felix, Michèle, Daniel and Max; sixteen grandchildren; five great-great grandchildren

EISENSTEIN, SERGEI
Born: Jan 22 1898, Riga, Latvia
Died: Feb 11 1948, Moscow
Director

Russian director; he combines propaganda with visceral cinema in the screen classics *Strike* (1924), *Battleship Potemkin* (1925), *October* (1927), *Ivan the Terrible Part 1* (1945) and *Ivan the Terrible Part 2* (posthumously released in 1958); the pioneer of the montage technique.

John Ellis (picture courtesy of John Ellis)

ELLIS, JOHN
Born: Aug 31 1968, Crumpsall, Manchester
Pianist, singer, producer, Limefield label chief

Jazz pianist/singer from Manchester; as a performer and producer Ellis is as busy as the proverbial bee mascot of his home city; begins playing semi-pro at 14; studies at Leeds College of Music; groups include John Ellis Big Bang, The Cinematic Orchestra, Honeyfeet, The Breath, Baked a la Ska; founds Limefield Studio at his home in Middleton, 2005; coaxes near neighbour Bill Leader out of retirement; between 2011-2015, their joint productions include – various: *Oddfellows*; Ian Reynolds, *Shreds*; Desi Friel, *The Knowing of You*; Dead Belgian, *Love and Death*; Rioghnach Connolly, *Black Lung*; Hunter Muskett, *That Was Then, This Is Now*; Matt Owens, *The Aviators' Ball*; Trevor Hyett, *Eager & Anxious.*

FIELDS, GRACIE
Born: Jan 9 1898, Rochdale
Died: Sept 27 1979, Isle of Capri
Singer and actor

Rochdale's finest, a working-class lass of relentless cheerfulness and vivacity, at her peak in that strange decade, the thirties, when there was an overwhelming need for cheerfulness, vivacity and belief. Whereas modern audiences might find her brassy, hyper and flat in voice, she was nothing less than a divinity to her original fan-base. Now we might suppose that the full employment of shipyard workers on the Clyde in 1939 had to do with the advent of war, whereas her original fan-base knew it was 'our Gracie' that did it, citing *Shipyard Sally* as evidence; that's the one with 'Wish Me Luck As You Wave Me Goodbye'.

FIRESTEIN, JACK
Born: 1917, Whitechapel, London
Died: 2004, London
Anti-fascist, bookseller, street fighter, hero of Cable Street and Anzio, director of Unity Theatre

One of six children of Jewish immigrants from East Europe, Jack follows his father in the tailoring trade before turning bookseller – beginning as co-proprietor with brother Phil of Carters, a communist bookshop; joins the CPGB, circa 1930; a veteran of the Battle of Cable Street, where East End residents rout Oswald Mosley's Blackshirts, 1936; as one of the 'D-Day Dodgers', he is injured at the Battle of Anzio, 1944; charms his Italian nurses and persuades a German SS man that Firestein is a good English name; runs a bookstall at Unity and manages the skiffle night and folk club there, 1959-1975; manages a Unity touring group to entertain at left-wing gatherings – "we went out 'on mobile'" says John Foreman, who explains why some Unity people disapproved of Jack: "they were all there looking for women or looking for men or looking for this or looking for that and Jack used to scoop the pool with the ladies"; personal chauffeur to trade union leader Clive Jenkins; his book stall is a Camden Town fixture on Saturdays, sited opposite the Camden Labour Party offices; a trustee of the Camden Neighbourhood Advice Centre; joins George Galloway's Respect Party, 2004; in a throwback to '44, he entertains nurses with renditions of Red Army and socialist songs on his deathbed; the subject of *Only a Bookseller*, a documentary film by Chris Reeves, 2009.

John Foreman (picture by Eva Navarro)

FOREMAN, JOHN
Born: Jan 14 1931, Euston, North London
Singer, activist, music hall populariser, printmaker, a keeper of ephemera; the Broadsheet King

Forebears include a circus ringmaster (albeit "Uncle Charlie ended as a street photographer") and music hall artistes Elsie Naish and Victoria Lytton; champions music hall when it is unfashionable and unprofitable; host of the skiffle night at Unity Theatre as the mainstay of Smoky City Skiffle group, 1956 onwards; Mr Jingle in *The Misadventures of Mr Pickwick*, a musical by Arnold Hinchcliffe, Unity Theatre, 1960; designs and prints programmes for Unity inc. *Bloomsday* (Michael Gambon debut), 1960; turns out the *Streets of Song* booklet (12T41) and other inserts for Topic Records, plus *Make Your Own Beer* (see *DALLAS, GLORIA*) on Unity's Gestetner duplicator; produces songsheets as the Broadsheet King, advertising his wares by singing them at Petticoat Lane market on Sundays; produces a songbook for each year of the Aldermaston March, 1959-63; sings the songs he sells as sheets – inc. works by Ewan MacColl, Sydney Carter, Karl Dallas *et al* – at Petticoat Lane on Sundays; serves three months in Brixton Prison as a conscientious objector, 1953-1954; assistant doorman of the Metropolitan music hall, Edgware Road; bottles for Professor Alexander [see *ALEXANDER, JOHN*); an honorary member of The Alberts, Bruce Lacey's group of madcap Situationists; produces *In Memoriam* for Winston Churchill's Lying in State, 1965; a supply teacher; republishes Charles Hindley's broadsheet collection *Curiosities of Street Literature*, 1966; retires to care for his mother; the subject of *John's Songs – A Day with a Music Hall Master*, BBC Radio 4 (2016, repeated 2017).

Recordings: *The 'Ouses in Between*, Reality RY 1004, 1966; *Folk Scene*, FSP 001, 1966 ('The Ratcatcher's Daughter'); *The Londoners*, Folktrax JFTX 331, 1975 (a cassette of Peter Kennedy recordings also featuring Fred Luckhurst, Bill Burnham and Bill French); *From Liszt to Music Hall*, Open University OU45, 1978 (shares side two w. Harry Boardman; Liszt occupies side one).

Personal: m. Rita (?-1970, sep.); two sons, Nat and Chris, the latter is guitarist with Madness.

GAIMON, BRIDGET
Born: circa 1844, County Clare?
Died: Dec 21 1916, Leytonstone, London
Street trader, Bill and Jo's great grandmother.

Born Bridget Ryan in (we think), County Clare; married Patrick Leader (see *LEADER, PATRICK, 1*), 1864; Patrick d. 1877; m. Michael Gaimon, 1879, though she was always known as 'Great Granny Leader' or 'old Mrs Leader'; listed in the 1891 census as a greengrocer in Rathbone Street, a street market in Canning Town.

GALVIN, PATRICK
Born: Aug 15 1927, Cork
Died: May 10 2011, Cork
Poet, singer, playwright and author

A childhood marked by chronic poverty – two parents and seven children in a two-roomed attic flat in Margaret Street, Cork; leaves school at 11 to work full-time, equipped with a false birth certificate and barely able to read and write; nevertheless he sells ballads and recites in public bars, and regrets that bards no longer have a practicable career path; sent to Daingean Reformatory School in Co. Offaly, notorious for its abusive regime; joins the RAF, still under-age, 1943; co-founds and edits the literary magazine *Chanticleer*, 1952-54; taken up by uilleann piper Séamus Ennis after he hears him sing some Irish traditional songs on *As I Roved Out*; publishes *Irish Songs of Resistance*, a history of Ireland through song, a bestseller for the WMA, 1955; *Irish Drinking Songs* (Riverside RLP 12-604) has arrangements by Will Sahnow, 1956; *Heart of Grace*, his first poetry collection, 1957; *And Him Stretched*, his first play, produced by Unity Theatre, 1961; the Archbishop of Dublin commissions a report about Galvin's 1962 play, *Cry the Believers*, concluding, "With its unending carping about the Church and the clergy, it was not a play to which young, impressionable minds could be exposed without risk to faith"; his *Raggy Boy* trilogy – *Song for a Poor Boy* (1990), *Song for a Raggy Boy* (1991) and *Song for a Fly Boy* (1999) – anticipates and improves upon the misery memoir; writer in residence at Dún Laoghaire in Dublin in the early nineties; a stroke confines him to a wheelchair, 2003; celebrates Cork as City of Culture with *Everything But You*, a translation of the works of Turkish poet Yilmaz Odabashi, 2005.

GOLDSBROUGH, ARNOLD

Born: Oct 26 1892, Gomersal, W. Yorkshire
Died: Dec 14 1964, Tenbury Wells, Worcestershire
Organist, harpsichordist, conductor and teacher

Born in Gomersal, nr Bradford, and recognised early as a musical prodigy; assistant organist at Manchester Cathedral, 1917; organist at St Anne's, Soho, 1920-3; assistant organist at Westminster Abbey, 1920-7; teacher at Royal College of Music, from 1923; succeeds Gustav Holst as director of music, Morley College, 1924–9; organist at St Martin-in-the-Fields, 1924-35; founds Goldsbrough Orchestra, 1948, to popularise less familiar composers, inc. Locke, Scütz, Lassus and Rameau; renamed the English Consort Orchestra, 1960; a pioneer of the early music movement.

GOLDSTEIN, KENNETH S.
Born: Mar 17 1927, Brooklyn
Died: Nov 11 1995, Philadelphia
Folklorist, producer, mover and shaker

A degree in business studies from the City College of New York and army service in the forties; impresses Stinson, Folkways, Prestige and Riverside Records sufficiently for them to entrust him with A and R; records Jean Ritchie, Rev. Gary Davis, Lightnin' Hopkins, Sonny Terry and Brownie McGhee a.o.; crucially, he spots a gap in the market for English folk in the US, and contacts the WMA to act as agents, 1956; founder of Philadelphia Folk Festival, 1962; Program Director of Philadelphia Folk Festival, 1962-1977; gains the first Ph.D. in folklore from the University of Pennsylvania, 1963; develops methodology for a generation of folklorists as professor at UPenn and author of *A Guide for Fieldworkers in Folklore*, 1964; secretary and treasurer of the American Folklore Society, 1966-72; head of the Dept of Folklore at Memorial University of Newfoundland, 1976-78; estimated to have produced over 800 albums.

Further reading: *Fields of Folklore: Essays in Honor of Kenneth S. Goldstein*, ed. Roger D. Abrahams, Trickster Press, 1995.

GRANEY, PAUL
Born: Aug 3 1908, Pendle, Lancashire
Died: Jun 1982, Manchester
Song collector, pipe fitter, activist

A 12-year-old 'half-timer' (a boy who spends half the day at school and half working in the mill); aged 18, responsible for picket action in Trafford Park during the General Strike; unemployed after a string of dead-end jobs and too proud to live off the meagre earnings of his mother and sister, Graney joins the ranks of the homeless and criss-crosses the UK moving from one doss-ouse ('spike') to the next, *c*. 1928-30; an organiser in the National Unemployed Workers' Movement, he is involved in the Battle of Bexley Square, where police attack a demonstration by unemployed workers, Salford, October 1, 1931; a charter member of the Youth Hostel Association, circa early 30s; active in the Mass Trespass protests, 1932; organises the

North West contingent of the Jarrow March, 1936; visits Holland and Germany and witnesses a Nazi rally in Munich, 1934; a stagehand at Joan Littlewood and Ewan MacColl's Theatre of Action, Manchester, 1934; an active song collector and avid patron of folk clubs; revival singers with cause to be grateful for his generosity inc. the Watersons, Harry Boardman, Marie Little, Mike Harding, Bernard Wrigley, Lea Nicholson etc; he amasses a huge library of tapes – radio documentaries, private live recordings, oral histories, a spoken autobiography, aka 'the memory tapes' – in an archive presently stored at Manchester Central Library; a lifelong photographer, strong on portraits and industrial landscapes.

Book: *One Bloke – A Manchester Man's Tale of Two Decades*, ed. Barry Seddon, Bluecoat Press, 2011.

Quote: "I never saw the whole picture, because only the intellectual who stands at the back and reads the books ever sees the whole picture. The bloke at the front getting his ribs kicked in sees nothing except the bloke who is kicking him."

Cy Grant (picture by Brian Shuel)

GRANT, CY
Born: Nov 8 1919, Beterverwagting, Guyana
Died: Feb 13 2010, London
Singer, actor, author, war hero

Receives a colonial education in British Guiana (strong on English history); a navigator in the RAF in the Second World War; shot down over the Netherlands; a POW in Stalag Luft III camp, 1943-1945; blocked in his chosen career of law by racism, he turns to entertainment and becomes a household name singing topical calypsos on BBC's weekday news and magazine programme *Tonight*, *c.* 1960; introduces 'Feeling Good' in the Newley and Bricusse musical *The Roar of the Greasepaint*; co-founder and director of the Drum Arts Centre in London, a centre for black artistic talent, 1974-1978; publishes poetry and philosophical memoirs.

Selected albums: *Folk Songs & Cool Songs*, Transatlantic TRA 108, 1962; *Cool Folk*, World Record Club, 1964; *Cy & I*, World Record Club, 1965; *Ballads, Birds & Blues*, Reality, RY 1006, 1966.

As an actor: *The Honey Pot*, 1967; *Shaft in Africa*, 1973; *At the Earth's Core*, 1976; the voice of Lieutenant Green in Gerry Anderson's *Captain Scarlet and the Mysterons*, 1967-68; the title-role of *Othello*, Phoenix Theatre, Leicester, 1965.

Personal: m. Dorith, four children.

GREEN, TONY
Born: Jun 25 1943, Birmingham
Folklorist, song collector, member of the Oral History Society

A contributor to Leeds University's folk fanzine *Abe's Folk Music* as a student (Bob Pegg is editor); a latter-day song collector, he donates his field recordings to the Leeds Archive of Vernacular Culture at Leeds University; founder member of the Oral History Society.

Quote: Oral historians need "to concentrate much more on history as what people think happened, including the presentation of radically different accounts, in order to demonstrate that different individuals and groups experience the same event in totally different ways, and to analyse why this is so" – Tony Green at the Oral History Society 1972 conference.

Martin Hall (photo by Roger Liptrot)

HALL, MARTIN
Born: May 2 1954, Bury, Lancs
Died: March 20, 2018, Isle of Skye
Accordion, concertina, guitar, banjo, vocal

A member of Jolly Jack, w. Alan Taylor and Dave Weatherall, a close-harmony group that graced the North West clubs in the eighties; Hall's sonorous voice, as "deep as a dungeon", was only infrequently heard at Oddfellows; his reticence made the impact the greater when he did sing; in truth, he feared his voice was much diminished due to underlying MS; Martin compensated by playing every instrument he picked up, and playing it very well, favouring accordion, concertina and guitar in Jolly Jack and banjo at Oddfellows; a gentle giant at 6ft 4ins.

Discography: with Jolly Jack – *Rolling Down to Old Maui* (Fellside, 1983); *Long Time Travelling* (Fellside, 1988); solo – *Ringing the Changes* (Fellside cassette 1992); with The Legplaiters, *The Years Behind the Times* (Cock Robin, 2002).

HALL, REG
Born: May 1933, Gravesend, Kent
Melodeon, piano; organiser, author, scholar and probation officer

Accompanies Irish fiddlers Michael Gorman and Jimmy Power in London pubs, circa mid-fifties; plays piano and melodeon in the dance-band The Rakes, 1956-present; plays for the Bampton Morris and the Padstow Blue Ribbon 'Obby 'Oss; with Mervyn Plunkett, discovers Scan Tester, Pop Maynard a.o.; with Mervyn Plunkett, launches the magazine *Ethnic* "for those who have the good of the English tradition at heart", 1959 – its impact is felt beyond its four issue lifespan; works with Bill Leader on *Paddy in the Smoke* (121T176), *Martin Byrnes* (LEA 2004), *Irish Dances* by Jimmy Power (BY 6040) a.o.; curates *The Voice of the People* series, 1999-2016.

Books: *I Never Played to Many Posh Dances – Scan Tester, Sussex musician, 1887-1972* (Music Traditions, 1990); *A Few Tunes of Good Music*, no longer available at http://www.topicrecords.co.uk/a-few-good-tunes/.

Awards: The EFDSS Gold Badge, 1987; the Gradam Cheoil musician's award from the Gaelic television company TG4, 2009.

Mike Harding (picture courtesy of Bill Leader)

HARDING, MIKE
Born: Oct 23, 1944, Crumpsall, Manchester
Comic, singer, author, playwright, broadcaster

After work on the buses and digging roads, plus reviving a pace-egging play still performed annually in Middleton, Harding makes his record debut on *Deep Lancashire*, Topic, 1968; his debut full-length LP is *A Lancashire Lad*, Trailer, 1972; 'The Rochdale Cowboy', his breakthrough, reaches #22 in the singles chart, 1975; a high profile career as comic, broadcaster and author follows; his literary output ranges from humour – *Napoleon's Retreat from Wigan, The 14½ Pound Budgie* – to erudite studies of folklore and art (*The Little Book* series), poetry (*The Singing Street, Cosmos Mariner*), and panegyrics to rambling (*Walking the Dales, Footloose in the West Of Ireland, Walking In The Peaks & Pennines*); a series of live double albums chart the eclipse of Harding the singer by Harding the stand-up – *One Man Show*, Philips, 1976, *Captain Paralytic & The Brown Ale Cowboy*, Philips, 1978, *Komic Kutz*, Philips, 1979; hosts *The Mike Harding Show*, BBC radio's flagship folk programme from 1997; ditched by orders from on high, 2012; the break with the BBC serves to unmuzzle the political commentator – a letter to Theresa May goes viral, 2018; only Billy Connolly rivals him among his immediate peer group – cultural icons to have crashed on Bill Leader's floor.

John Hasted (picture by Ernie Greenwood, courtesy of fRoots archive)

HASTED, JOHN
Born: Feb 17 1921, Woodbridge, Suffolk
Died: May 4 2002, Penzance, Cornwall
Physicist and champion of vernacular urban music, or skiffle and folk

Educated at Winchester College; wins chemistry *and* choral scholarships to Oxford; joins the Workers' Music Association, 1940; founder of the Oxford Workers' and Students' Choir, 1940; joins the Local Defence Volunteers and gains a Chemistry BA, Oxford, 1940; in the British army, installs radar in Malta and Alexandria and sees action in the Italian campaign, 1941-45; his theories on revolutionary music are confounded by a 78 by the Almanac Singers, 1946; joins Sir Harrie Massey's team at University College, London, 1948; assists Alan Bush directing the WMA Choir, 1948; forms The Ramblers w. Bert Lloyd, modelled on the Almanac Singers, 1950; founder of the London Youth Choir, 1951; faces down the US army at Innsbruck – they try to intimidate the LYC from continuing to Berlin for the Youth Festival, 1951; writes 'Go Home Yankee', his first politically-motivated song; co-founds *Sing* magazine and becomes its music editor, 1954-mid-sixties; a regular tutor at the WMA Summer School, Wortley Hall, South Yorkshire, 1954 onwards; co-writes 'The Ballad of Wortley Hall' with Bill Leader, 1954; teaches and popularises the guitar, keeping one step ahead with lessons via correspondence with Pete Seeger; musical director of Australian folk opera *Reedy River*, Unity Theatre, 1954-55; founder of The Good Earth, at 44 Gerrard Street, Soho – one the first folk clubs in the UK, 1954; Redd Sullivan, Martin Winsor, Shirley Collins becomes residents of the renamed Forty-Four Club, the nurse-bed of skiffle, 1954-1956; a policeman smashes his Martin guitar during an anti-Suez protest, 1956; visits Woody Guthrie, 1958; "steps in and does anything that is needed" for Topic, driving Bill to Saltaire for the Cait O' Sullivan/Alex Eaton sessions and playing piano for Bill Owen on 'First Things First' (abandoned); accompanies Dominic Behan for Topic and Shirley Collins for Argo; reproaches himself twice over for 1) encouraging the break-up of the LYC into small skiffle units and 2) giving up music to focus exclusively on science at the behest of second wife, Lynn; Professor of Experimental Physics and head of the physics department at Birkbeck College, 1968; his credibility is questioned when he publicly endorses Uri Geller's spoon bending feats; retires as professor, 1984.

Personal: m. Elizabeth Gregson (1947-1958, div.); twin daughters, Annie and Belinda (b. Sept 7 1949); m. Lynn Wynn-Harris (1959-1988, her death); one son, John Andrew (b. 1963).

Selected books: *The Physics of Atomic Collisions*, Butterworths, 1964; *Aqueous Dialectrics*, Chapman & Hall, 1973; *The Metal Benders*, Routledge, Kegan Paul, 1981; *Alternative Memoirs*, Greengates Press, 1992.

Quote: "If we accept what has always seemed more likely, namely that the universe behaves as a closed system, then we must be continually watchful for unexpected phenomena, that is to say, for miracles. It is such discrepancies which offer clues to any deficiencies in existing theory."

Hamish Henderson (picture by Brian Shuel)

HENDERSON, HAMISH

Born: Nov 11 1919, Blairgowrie, Perthshire

Died: Mar 8 2002, Edinburgh

Song collector, poet, songwriter, soldier, cultural and political activist, 'remembrancer'

Born to a single mum and raised in a cottage in Perthshire; inherits his love of song from her – she sings in Gaelic, Scots and French – and his love of justice from his Jacobite grandmother; wins a scholarship to Cambridge University, where he studies German and French and defends the Spanish Republic in debate; a Quaker group hire him to rescue Jews from Germany, 1939; an intelligence officer in the Western desert, Sicily and Italy, 1941; promoted to captain, 1943; personally accepts the surrender of Italy from Marshal Graziani, April 29, 1945; *Elegies For The Dead In Cyrenaica* wins the Somerset Maugham award for poetry, 1949; he doubles the £660 prize money by placing £10 on a 66-1 outsider in the Grand National; the beginnings of the Scottish folk revival can be traced to the tour guide to Edinburgh he gives Alan Lomax, 1951; with Calum MacLean, founds the School of Scottish Studies at Edinburgh University, 1951; with Norman Buchan, organises the Edinburgh People's Festival, forerunner of the Edinburgh Fringe, a riposte to the official Edinburgh Festival, 1951; with Maurice Blythman, he subsumes his energy into the Glesca Eskimo collective, with its anti-Polaris, pro-Republican agenda; *Ding Dong Dollar*, the Folkways LP, is their statement, 1962; 'discovers' Jeannie Robertson, Flora MacNeil, the Stewarts of Blair, Calum Johnston; a champion of gay rights – when Hugh MacDiarmid tells him "homosexuality has never been a Scottish thing", Henderson counters with "homosexuality is a natural human thing"; refuses an OBE from the Thatcher government and is voted Scot of the Year by listeners to Radio Scotland.

Songs: 'D-Day Dodgers', 'Farewell to Sicily', 'Freedom Come All Ye', 'Rivonia', 'The John Maclean March' a.o.

Further reading: *Hamish Henderson: A Biography Volume One – The Making of the Poet (1919-1953)*; *Hamish Henderson: A Biography Volume Two – Poetry Becomes People (1952-2002)* by Timothy Neat.

HIBBERT, GEOFFREY

Born: Jun 2 1922, Sculcoates, Hull
Died: Feb 3 1969, Epsom, Surrey
Film, stage and television actor

Makes his screen debut as the juvenile lead in *The Common Touch*, dir. John Baxter, 1941; stars in *Love on the Dole*, dir. John Baxter, 1941, *In*

Which We Serve, dir. Noel Coward and David Lean, 1942, *Albert, R.N.*, dir. Lewis Gilbert, 1953, *I Was Monty's Double*, dir. John Guillermin, 1958, a.o.; appears alongside Julie Andrews in the original Broadway production of *The Boyfriend*, The Royale Theatre, 1954-1955.

HITCHCOCK, ALFRED
Born: Aug 13 1899, Leytonstone, London
Died: Apr 29 1980, Los Angeles, California,
Director

The master of suspense. His life and works are too well-known for summary here.

HOBSBAWM, ERIC
Born: Jun 9 1917, Egypt
Died: Oct 1 2012, London
Historian

The preeminent Marxist chronicler of the twentieth century believed that politics held the key to truths as well as myth and that turbulent century could only be understood by those who lived it; a scholar at the Prinz Heinrich-Gymnasium in Berlin and Kings College, Cambridge – the transfer came as a result of the rise of the Nazis; joins the Communist Party at Cambridge, like many another; a double-starred first in History; serves in the Royal Engineers and the Army Educational Corps during the war; becomes a lecturer in history at Birkbeck College, 1948; signs a letter of protest against the Soviet invasion of Hungary though he remains in the Party, 1956; supports the Prague spring; calls Tony Blair "Thatcher in trousers".

Reading: writes about jazz as Francis Newton (*The Jazz Scene*, MacGibbon and Kee, 1959); his trilogy of books about "the long nineteenth century" – *The Age of Revolution, The Age of Capital: 1848–1875, The Age of Empire: 1875–1914* (and a successor about the short 20th century, *The Age of Extremes*) are key works.

Quote (on the lure of bandits): "They are not so much men who right wrongs, but avengers, and exerters of power; their appeal is not that of the agents of justice, but of men who prove that even the poor and weak can be terrible" – from *Bandits*, Pelican, 1972.

John Howarth (photo by Roger Liptrot)

HOWARTH, JOHN
Born: Feb 17 1943, Oldham, Lancashire
Singer, nurse, song collector

A champion of indigenous Lancashire song, children's song and dialect poetry; pre-Tinkers, his school group, The Gamblers, perform on the touring show of *Six-Five Special*, the Empire Theatre, Oldham, 1958; forms the Oldham Tinkers with Larry and Gerry Kearns after a casual encounter with old classmate Larry, circa mid-sixties; the group make their record debut on *Deep Lancashire* (12T188), 1968; six albums follow on Topic, 1971-1977; the Tinkers provide songs for BBC's *Play for Today – Kisses at 50* by Colin Welland, 1973; the stars of the Queen's Silver Jubilee Royal Gala Performance, Palace Theatre, Manchester, 1977; by profession a state-registered nurse with the NHS; a patron of the Edwin Waugh Dialect Society; a mainstay of the session at The Oddfellows Arms, Middleton.

Nic Jones (picture by Janet Kerr)

JONES, NIC
Born: Jan 9 1947, Orpington, Kent
Guitarist, fiddler, singer, inspiration

In his mastery of acoustic guitar techniques, handed down primarily from Martin Carthy and given a joyous spring all its own; in the utter commitment and conviction with which he exhumed the old bones of the tradition to create something fresh; and in his easy, outgoing personality, Jones is the all-round folk hero; a generous accompanist, everyone noticed how much better they played when Nic was about; his association with Bill resulted in four Trailer albums – *Ballads and Songs*, *Nic Jones*, *The Noah's Ark Trap*, *From the Devil to a Stranger* – and notable collaborations (*Bandoggs*, Songs of a Changing World with Jon Raven and Tony Rose); a Tony Engle produced successor, *Penguin Eggs*, is one of those rare, happy occasions where pop accessibility and folk integrity meet without loss on either side; canonisation followed, but not before a car crash on Feb. 28 1982 deprived him of his motor skills and took everything away but his spirit; the best-loved man in the business.

Personal: m. Julia Seymour, 1968; three children, Daniel, Helen and Joe.

Awards: The EFDSS Gold Badge, 2012; Folk Singer of the Year at the BBC Radio 2 Folk Awards, 2013; *Penguin Eggs* voted second place in the 'Best Folk Album of all Time' by listeners of the *Mike Harding Show*, 2001.

JUDD, MARY
Born: 1879
Died: ?

The wife of Bartlett Judd, a shop-owner at 155 Earls Court Road in the early 1900s – oil, colours, china and glass – takes a shine to shop assistant Lou, and they go to theosophy lectures and such; a suffragette and an enlightened employer; one son, Harold, b. 1901.

Stan Kelly (picture by Brian Shuel)

KELLY, STAN
Born: Sep 15 1929, Merseyside
Died: Apr 16 2014, Oswestry, Shropshire
Writer, songwriter, the first Computer Sciences graduate in the world

Born Stanley Bootle; father Arthur is reportedly the best plumber on Merseyside; a star student at Liverpool Institute; National Service in the Royal Electrical and Mechanical Engineers,1948-50; gains a scholarship to Downing College, Cambridge, reading science and mathematics; studies under Maurice Wilkes and attends a lecture by Alan Turing; a member of St. Lawrence Folk Song Society, the Cambridge folk club, with Rory McEwen and Leon Rosselson; an early graduate in computer science, 1954; works for IBM in the UK and USA, 1955-1970; Manager for University Systems for Sperry-UNIVAC, 1970-1973; a freelance computer consultant and writer, from 1973; he enlivens computer jargon with his linguistic flair whilst leading a double life as a folk singer; his song, 'Liverpool Lullaby' becomes a folk club standard, variously covered by the Three City Four, the Ian Campbell Folk Group, Judy Collins, Cilla Black, a.o.; an avid supporter of Liverpool FC; anticipates the information age with the phone service, Soccer-Dial, too ahead of its time to succeed; returns to the UK in frail health, cared for by artist daughter Michele Coxon, 2004-14.

Personal: m. Peggy Jones, 1948-75, div.; five children, Edmund, David, Anna (d. 1985), Carol, Michele; two daughters, Cressida and Kate, from later relationships.

Songs: 'Liverpool Lullaby', 'I Wish I Was Back in Liverpool' (w Leon Rosselson), 'My Old Man' aka 'What Was the Colour', 'Last Night We Had a Do', 'The Quality of Mersey', 'Romeo and Juliet', 'The Ould Mark II', a.o.

Selected discography: *Liverpool Packet*, TOP27, 1958; *Songs For Swinging Landlords To*, TOP60, 1961; *I Chose Friden – Songs for Cybernetic Lovers*, private pressing, 1963; *O Liverpool We Love You*, Xtra, 1969; *Echoes of Merseyside*, Liverpool Echo, 1971, a tribute to the city of Liverpool.

Selected books: *Liverpool Lullabies, The Stan Kelly Songbook*, pub. Sing, 1960; *Lern Yourself Scouse – How to Talk Proper in Liverpool*, pub. Scouse Press, 1961; as Stan Kelly-Bootle – *The Devil's DP Dictionary*, pub. McGraw-Hill, 1981; *680x0 Programming by Example*, pub. Howard W. Sams, 1988; *Computer Contradictionary*, pub. MIT Press, 1995.

Andy Kenna (picture courtesy of Bill Leader)

KENNA, ANDY
Born: June 24 1944, Liverpool
Singer, musician (concertina), teacher

Exudes the heartiness and bluffness of a true son of Liverpool, and is immersed in maritime music; performs as a youth in the family band, The Shannon Star Ceilidh Band; runs clubs in Liverpool and Manchester during the early revival; a charter member of The Mersey Shantymen, backing the likes of Bert Lloyd and Stan Hughill; emigrates to Nova Scotia, Canada, 1969; upon returning, settles in Bacup,

Lancashire; forms the Bury Ceilidh Band, later to become The Valley Band, a Lancashire dance group; performs at major maritime festivals throughout Europe, sometimes as half of 'Liverpool Forebitter', 1995 onwards; the resident 'Scouser' at the Monday night session at The Oddfellows Arms in Middleton, 2000-present.

Albums: (as Liverpool Forebitter) *Liverpool Forebitter*, *Saltwater Ballads*; (and solo) *Emigrants Served*, *Old Ports of Call – The Maritime Songs of Bob Watson.*

KINSEY, RAY

Born: Mar 5 1920, Bradford, West Yorkshire

Died: 1997

Boy soprano, squadron pilot, documentary film-maker, Livingston Studios chief

Joins the choir of St Philip's, Girlington, Bradford, aged 8; as boy soprano, a silver cup at the Blackpool Festival in 1932 brings him to the attention of HMV; makes six 78s for the label between January and September 1933; a squadron pilot; invalided out of bombing raids by stress disorder; appears on children's TV (*Sunday Special – The Needs of Others*) showing his pictures of Chinese refugees in Hong Kong, Apr 27, 1958; with Fred Livingston–Hogg, founds Livingston Studios in Barnet, originally to dub sound onto film, 1963; the diversification to music, particularly folk music, gathers pace when son Nic joins the team and Bill Leader becomes a regular client, 1966 onwards; retires as Livingson's manager when the studio moves to new premises in 1980.

Filmography: for Raymond Kinsey Productions, as director – *Lambeth 1958*, 1958; *The Present Challenge in Central Africa*, 1959; *The Seeing Hands*, 1964; *Books in the Making*, 1964; as producer only – *Treasure in India*, 1964; as producer and writer – *Nigerian Pattern*, 1954; for Livingston Recordings, as director and producer – *No Lack of Courage*, 1968; as producer – *New Day*, 1965; *John Groom's*, 1966; *This is the Church Army*, 1967; as producer and commentator – *Spring in Iran*, 1966

LEADER, EDWARD (TED)

Born: Sept 25 1908

Died: 1983

Lorry driver

"We didn't see a lot of Ted, and nothing Ted said or did gained approval." That's Bill, summarising his mother's attitude to her brother; his cousin Jo, on the other hand, says that most of the family got on with him reasonably well ("I think they were happy with him most of the time"), and adds that all the Leaders of Ted's generation had a temper, even her sweet-natured father, Patrick, who made himself ill by suppressing it.

Personal: m. Violet, 1938; one son, Edward, b. 1942.

Helen Leader (with John Doonan and Bill at the opening of the Irish Eating House, 1968; picture by Peter Trulack)

LEADER, HELEN
Born: Oct 6 1943, Anderston, Glasgow
Activist, organiser, co-founder of Leader, Trailer, Hill and Dale

Born Helen Roberta Heron; a political activist since the anti-Polaris demonstrations, Holy Loch, 1959; follows husband Bobby Campbell to London, 1964; works as a secretary in the advertising dept of the *Daily Worker*, circa mid-sixties; works in the offices of The Russian Shop, London; founds, with Bill Leader, the Leader and Trailer labels, and takes care of the business, 1969-78; ditto for a third independent label, Hill and Dale, 1978-82; volunteer and then organiser on the founding committee of the Halifax Well Woman's Centre, opens 1984; organiser/administrator of Bannerworks, a community arts group dedicated to the preservation of trade union banners; works at Pecket Well College, a user-run, managed collective for people with learning difficulties, circa mid-nineties-2003; a Greenham Common veteran.

Personal: m. Bobby Campbell (1963-1966 sep.); m. Bill Leader (1969-1993 sep.); daughter Amy, b. 1977.

Lou Leader in the garden of 5 North Villas (picture by Bill Leader)

LEADER, LOUISA JANE (LOU)
Born: Aug 9 1896, Canning Town, London
Died: Mar 20 1974, Greetland, West Yorkshire.
Activist, factory worker, Bill's mother

Born Louisa Leader; educated at St Margaret's, Canning Town; goes into service and is befriended by Mary Judd, the wife of her employer; emigrates to the USA, 1924; works as a nanny for Judge Coleman, Baltimore, circa 1926-28; marries her cousin, William Leader, Mar 24 1928; lives in New Jersey where son Bill is born Dec 26 1929; the family return to the UK, 1931; Becontree, 1933-37; Mottingham, 1937-1941; the move to Yorkshire follows the evacuation of Smart & Brown, where Bill Snr works, to Keighley, 1941; Saltaire, 1941-1960; a wartime factory worker at Parkinson's Perfect Vices, Shipley; contributes self-made rugs and woollen toys to the *Daily Worker* annual bazaar, to raise funds for Britain's Communist daily newspaper; moves to North Villas, Camden Town, 1960; widowed 1961; moves to Greetland with Bill and Helen, 1974; dies Mar 20 1974.

"Her tongue was crisp in defence of the Party and its ideals as she fought for better working conditions and equality for women in the factories. Louisa was outstanding, but she was one of many" –Alex Eaton.

Lynne Leader (picture courtesy of Bill Leader)

LEADER, LYNNE
Born: March 5 1960, Lowestoft
Librarian

Née Porter. Acquires an informal music education at the Record Exchange in Lowestoft; moves to Manchester in 1979 as student in Library and Information Studies; works at University of Liverpool 1982-85 before moving to Salford as Music Librarian, where, as Lynne Sharma, she meets Bill; plans and develops a new music, media and performing arts library at Adelphi Building in 1992; works as Head of Collections prior to retirement in 2018.

Personal: m. Chandra Sharma (1985-87, sep.); daughter Annie, b. 1993; m. Bill Leader, 1998.

LEADER, MARGARET (MAGGIE)
Born: Aug 15 1903, London
Died: April 12 1992, Sheffield
Tailoress, Bill's cousin Jo Pye's mother

Née Wicks, marries Patrick Leader, Lou's brother, March 8, 1941; acquires a regard for Jewry as a cleaner for the local fishmongers and Mr Schenke, the local rag and bone merchant; Mrs Schenke teaches Maggie odd bits of Yiddish, and the wherefores of kosher food and Sabbath rituals; her workmates and friends from Lotries tailoring firm (Whitechapel, later Aldgate) are Jewish; in an early job she is employed by Kealey & Tonge Ltd, makers of jams and pickles; she never eats any product by Kealey & Tonge after cleaning the vats and being ordered to leave the tin tacks and dead rats where she found them – *their weight was keeping them at the bottom*; via Jo, our witness to the South Hallsville School bombing, surveillance at 13 Park Grove, and the fostering of José Luis Tovar González.

LEADER, PATRICK
Born: circa 1839
Died: December 2 1877, East Ham
Stoker

An emigre from Ireland, with his brother Thomas, circa 1855; works as a stoker at Beckton Gasworks; sires Lou's wing of the Leader clan – he is Lou's grandfather.

At Hoddesdon (l. to r.): Maggie Leader, Jo Pye (née Leader), Bill, Bill Snr (standing), Lou, Patrick (Pat) Leader (4) (picture courtesy of Bill Leader)

LEADER, PATRICK
Born: Aug 1 1903, London
Died: Feb 22 1980, Bishop's Stortford, Herts
Merchant Navy, 1924-Oct 1939, heating & ventilation fitter

Auburn-haired and known as 'Ginger Pat' to distinguish him from all the other Pats in the family; the oldest of Lou's surviving brothers and Jo's father; at sea in the Merchant Navy when the Second World War breaks out; fits the ventilation at Bingley Teacher Training College, a long-term contract, and moves Maggie and Jo to Saltaire to stay with Bill Snr and Lou, 1942; unemployed in his youth; rumoured to have been sacked from Silver's in Canning Town when Lou tried to organise a trade union branch; rumoured, with sister Molly, to have followed Lou as an employee of Bartlett Judd's china shop in Chelsea (see *JUDD, MARY*) and to have been sacked for breaking the goods, probably groundlessly; "you know we Leaders all like a good story" (Jo).

Personal: m. Maggie, Mar 8, 1941; daughter Jo b. Aug 23 1942.

LEADER, THOMAS
Born: circa 1847
Died: between 1902 and 1911
Stoker

An emigre from Ireland, with his brother Patrick; circa 1855; works as a stoker; sires Bill Snr's wing of the Leader clan – he is Bill Snr's father.

LEADER, THOMAS
Born: Jul 15 1865
Died: Aug 25 1942
Stoker

A gas stoker at Beckton Gas Works; employed by the council, possibly as a road sweeper; lives in Canning Town and is the family patriarch; Bill and Jo Pye's shared grandfather.

Personal: m. Louisa Martha, Christmas Day, 1889; sons and daughters – Walter Thomas, Nance, Thomas, Lou, Molly, Harry, Patrick, Lily, Ted, Jim, Kit.

Quote: "My grandfather on my mother's side spent most of his time working in Beckton Gas Works. You were lucky to get out of one of those places alive" – Bill Leader.

LEADER, WILLIAM
Born: July 7 1888, London
Died: July 20 1961, Archway Hospital, London
Toolmaker, activist, Bill's father

Identified as Bill Snr for the purposes of this book, he was always 'Bill' in life; a skilled machine toolmaker, follows his cousin Lou to the USA; they marry in New Jersey Mar 24 1928; the family returns to the UK, 1931; work at Ford Dagenham, Vickers-Armstrong, Smart & Brown (machine tools and precision engineers); the family move to Yorkshire when Smart & Brown evacuate to Keighley to share the site of The Rustless Iron Company (TRICO), 1941; a lifelong trade unionist and activist; the Keighley branch of the Communist Party organise two coach-loads of mourners for his funeral at Golders Green Cemetery, July, 1961.

Jack Lee (photo by Roger Liptrot)

LEE, JACK
Born: Jun 13 1926, Rochdale
Died: Jan 18 2015, Rochdale
Singer, songwriter, monologuist, promoter, engineer

Jack and Mavis (1930-2017) go together; like some Tommy Cooper of folk, Jack has a hilarious way of forgetting lines; "What's next, Mavis?" become a catchphrase; self-penned monologues include 'George and the Dragon' and 'The Phantom of Balderstone Mill', based on a true story they say; as a long-standing promoter, operating from a phone box in the beginning, he brings 16-year-old Barbara Dickson to The Fisherman's Inn at Hollingworth Lake for her North

West debut in England; declines the offer of a reciprocal engagement at Billy Connolly and Gerry Rafferty's folk club; Mavis says, "I'm not going to Glasgow! We've only got a motorbike and a sidecar!"; his C&W club, The Hobo's Retreat, is a fixture for over four decades at various venues around the Heywood area; he and Mavis share the same birthday.

LEVIN, ARIEL
Born: 1908, Edmonton, Hertfordshire
Killed: Aragon, Spain 1938
Peace organiser

Quote: "I knew Levin, as it seemed everyone called him, because of my mother's involvement in the Barking Peace Council in the mid 1930s. He was the inspirational prime mover in the organisation, involving a wide spectrum of people, and originating many imaginative events, during this period of a coup in Spain, aggression in China and the threat of war in Europe. He impressed this seven-year-old boy as an inspirational and charismatic person" – Bill Leader.

LEVIN, DEANA
Born: Jul 15 1906, Hackney, London
Died: Mar 30 1980, Camden Town, London
Teacher and author

Her experience as a teacher in a school for foreigners in Moscow provides raw material for much cited academic studies *Children in Soviet Russia* (Faber and Faber, 1942), *Soviet Education Today* (Staples Press, 1959), *Leisure and Pleasure of Soviet Children* (MacGibbon and Kee, 1966), and a children's book, *Nikolai Lives in Moscow* (Hastings House, 1968).

Quote: "I can't remember how I first got to know of Deana Levin, Ariel's sister, but I was aware of her name from childhood. I met her just once and briefly. It was on a visit to Gerry Sharp's house in Nassington Road, Hampstead. It was a flying visit on my part and I wasn't expecting to come across a gathering of the local left-wing elite" – Bill Leader.

Martin Lynott (picture courtesy of Bill Leader)

LYNOTT, MARTIN
Born: Nov 29 1938, Moston, Manchester
Fiddler, teacher, maestro

At De La Salle Teacher Training College in Alkrington, Lancashire, joins The Beggarmen, with Tony Kelly, Terry Walsh and Norman Stainthorpe, 1966; Irishmen Gerry Brady and Eamonn Clinch complete the line-up; *The Beggarmen* (Studio Republic) is recorded live at The Beggarmen's club at The Crown and Anchor, Stevenson Square, and the Manchester Sports Guild, 1969; with Tony Downes, records *Flowers of Manchester* (Sweet Folk and Country) as The Two Beggarmen, 1979; at about this time (late seventies), Martin starts a folk club with the format of tune, song, song, song, tune; the club/session moves to many venues – The Barbers Arms, The Joiners Arms, The Ring o' Bells, The Old Boars Head, The Carters Arms – before settling at The Oddfellows' Arms on Oldham Road, Middleton, where they make their debut on December 22, 2008; Bill Leader is a regular; the nucleus of the Oddies crew – Lynott, John Howarth, Steve Keene, Andy Kenna, Martin Hall, Pete Macmillan, Mike Canavan, Ian Sidebotham, Dave Howard – form the Legplaiters

Ceilidh Band; Martin relinquishes the leadership role when he becomes ill with Parkinson's Disease, 2019.

Recordings: *Oddfellows – Monday Night at Nine* (2008, Limefield); *Oddfellows* (Limefield, 2011); as The Legplaiters Ceilidh Band – *The Years Behind the Times* (Cock Robin, 2002), *Another Year On* (Cock Robin, 2003).

Quote: "Thursday Night in The Crown and Anchor with The Beggarmen was the best gig in town back in 1966/7. Eamon and Gerry at the front on lead vocals, bones and spoons… Martin, Norman, Tony and Terry weaving a tapestry behind… The room was always crammed tight, the landlord seldom smiled, the ale was good, the crack was ninety. They gave me a few gigs when I was starting out. I heard Luke Kelly there. No PA, no lights, no production, just heartfelt good vibrations. The good old days were the days when we didn't talk about the good old days" – Christy Moore.

MARX, KARL
Born: May 5 1818, Trier, Germany
Died: Mar 17 1883, London
Revolutionary and economic philosopher

Proposes practical ways to make life better for ordinary people with *The Communist Manifesto*; thereafter his name becomes an 'ism' and a term of general abuse. Sean Davies reckons Karl Popper's *The Open Society and Its Enemies* demolishes Marx. The book was resting on the arm of the chair when I went to interview him, but I haven't so far caught up with it.

MINCHINTON, JOHN
Born: 1926, Deptford
Died: 2017
Film subtitler

An interest in East European cinema leads him to found Films of Poland with Peter Brinson (see *BRINSON, PETER*), 1953; here he becomes expert in subtitling foreign films in English; strikes out as an independent subtitler the year BBC2 opens, 1964; subtitles over 1,600 films, averaging more than one feature film a week; one of two people responsible for subtitling foreign films for Channel 4, 1982 onwards; retires at the age of 88 in 2014.

Personal: m. Doris Mead (1947-2014, her death), honeymoon in Czechoslovakia; two children, Ruth and Paul.

MONTEUX, PIERRE
Born: April 4 1875, Paris, France
Died: July 1 1964, Hancock, Maine, United States
Conductor

First viola and then conductor of the Concerts Colonne, accompanist Diaghilev's Ballets Russes; conducts the premieres of Debussy's *Prélude à l'après-midi d'un faun*, 1912, Ravel's *Daphnis et Chloé,* 1912, and, famously causing a riot, Stravinsky's *Le sacre du printemps*, 1913; as a naturalised American citizen (1942) conducts Metropolitan Opera, Boston Symphony Orchestra, San Francisco Symphony Orchestra, *et al*; significantly (for our purposes), he undertakes a short concert season in London with the LSO and rehearses at St Pancras Town Hall, June 1958.

MOSS, HENRY
Born: Apr 26 1954, Harpurhey, Lancs
Insurance-man, potman

Works in the Civil Service and other office jobs, 1973-1990; works in an insurance office in Manchester, ? – present; never happier than when Paddy Moloney of The Chieftains responds to an encore request with, "Disgraceful, no homes to go to, I suppose"; supplies musicians with beer, collects empties and prompts as required at the Oddfellows Arms, Middleton; a wide and voracious reader.

Eva Navarro (picture by Mike Butler)

NAVARRO LÓPEZ, EVA MARÍA
Born: Sep 2 1973, Alicante, Spain; Mexican at heart
Scientist, educator, speaker, artist, philosopher

Scientist, technologist and educator in computer science, control engineering, mathematics, artificial intelligence, complex networks, neuroscience, and more; shadows the footsteps of Alan Turing in Manchester, 2008-2020, and Santiago Ramón y Cajal in Madrid, 2016-17. Currently, she is a Reader in Data Science in the University of Wolverhampton, were she is the founder and director of the AiDAs Research Lab, dedicated to artificial intelligence and data science.

Personal: m. Mike Butler, Aug 25 2016.

OWEN, BILL
Born: March 14 1914, Acton, London
Died: July 12 1999, Westminster, London
Actor

Born William John Owen Rowbotham in Acton, London; an actor under contract to Rank; adapts *The Ragged Trousered Philanthropists* for Unity Theatre,1949; other Unity productions include *Burlesque*, with a part for Smoky City Skiffle, 1957 (see *FOREMAN, JOHN*); co-writes the lyrics for 'Marianne', a hit for Cliff Richard, 1968; becomes a household name as Compo in the BBC sitcom *Last of the Summer Wine*, 1973-2000.

Selected filmography: *The Way to the Stars*, 1945; *Dancing With Crime*, 1947 (as Bill Rowbotham); *When the Bough Breaks*, 1947; *Daybreak*, 1948; *Once a Jolly Swagman*, 1949; *The Ship That Died of Shame*, 1955; *Carry On Sergeant*, 1958; *Carry on Regardless*, 1961; *Georgy Girl*, 1966; *O Lucky Man!*, 1973.

POLLITT, HARRY

Born: Nov 22 1890, Lancashire
Died: Jun 27 1960, on the SS Orion, coming back from Australia
General Secretary of the CPGB from 1929-39, 1941-56

From a long line of Lancashire working class radicals; a `half-timer' at 12; apprenticed as a boilermaker in Manchester; joined the Independent Labour Party, 1908; led a shipbuilding workers' strike in Southampton, 1917; London secretary of the boilermakers, 1919; organiser for the Red International of Labour Unions, 1921; national organiser of the Communist Party of Great Britain, 1923; elected General Secretary of the CPGB, 1929; architect of 'The British Road to Socialism'; the subject of 'The Ballad of Harry Pollitt' aka 'Harry was a Bolshie' by Elin Williams.

Personal: m. Marjorie Brewer, 1925.

PORTMAN, ERIC

Born: July 13 1901, Halifax, Yorkshire
Died: December 7 1969, St Veep, Cornwall, England
Actor

Impressed by his raging anti-semitism, von Ribbentrop promises Portman his own film studio in Berlin once Germany wins the war (the German ambassador and Yorkshire-born actor are lunching together in 1938); the story gives a new slant to Portman's portrayal of a Nazi in *49th Parallel*; he also makes *One of Our Aircraft is Missing* and *A Canterbury Tale* for the Archers, films that transcend their origins as wartime propaganda; one of 17 Number Twos in cult TV series, *The Prisoner*, 1967-68.

Jo Pye (picture courtesy of Bill Leader)

PYE, JO
Born: Aug 23 1942, Shipley
Teacher, Lady Mayoress, amateur historian, artist

Née Leader; the daughter of Patrick and Margaret Leader; born in Shipley, baptised in Canning Town and schooled in Saltaire, London and Hertfordshire; she moves back and forth between Yorkshire and East London throughout childhood; chooses Bingley College for her teacher training course to be near "Aunt Lou and Uncle Bill" just as Lou and Bill Snr move to Camden Town to be near Bill and Gloria and grandson Tom, 1960; to Hoddesdon, Hertfordshire, 1953; to Sheffield, 1980; Lady Mayoress and consort of "the only Cockney Mayor of Sheffield" (Mike Pye), 2004-2005; *de facto* Leader family historian.

Personal: m. Mike Pye, January 6 1973; two daughters, Mikhaila and Louisa; one granddaughter, Rosie Milward.

Quotes: "She has a better memory than me about who's done what" – Bill Leader

Russell Quaye; Hylda Sims foreground (picture by Brian Shuel)

QUAYE, RUSSELL
Born: December 26 1920, Bromley
Died: 1984
Artist, skiffler, WWII survivor, teacher

His youth scarred by tragedy, Quaye lies about his age to join the RAF at 15; a rear gunner in a Lancaster bomber; travelling to Malta, his ship is torpedoed and he is left clinging to the wreckage; the healing power of art is tested when he enrols at Beckenham School of Art and forms skiffle group the City Ramblers – w. partner Hylda Sims and housemate John Pilgrim; sings and plays kazoo and quattro, a four-string guitar from Portugal, with the Ramblers; known as 'the jazz painter' – Pearl Bailey and Big Bill Broonzy sit for portraits on the recommendation of Max Jones of *Melody Maker*; he makes a pair of Scott Joplin-themed LPs with new partner, Bulgarian pianist Mimi Daniel *c.* 1973; dies following heart surgery, 1984.

RAAB, CAMILLA
Born: May 20 1922
Died: Jun 11 2004, London
Clerical administrator of the Workers' Music Association, editor

Née Betbeder; a star pupil at the Skinners' Company's School for Girls at Stamford Hill; studies at Bedford College, London University, the first higher education college for women in the UK; clerical administrator of the Workers' Music Association, 1953-1957; a keen traveller, she is in Hungary in 1956, the year of the uprising, and is involved in the anti-apartheid struggle in South Africa; emigrates to Australia and works at the Melbourne University Press, 1957; returning to England, she works at a publishers and steers Eric Partridge's *Dictionary of Slang and Unconventional English* to its 8th edition, 1984; resident in Hampstead Garden Suburb; a force in the Residents' Association and the community magazine, *Suburb News*.

Personal: m. Felix Raab, 1957-1962, his death.

RAMELSON, BERT
Born: Mar 22 1910, Cherkassky, Ukraine
Died: Apr 13 1994, London
Activist

Born Baruch Ramilevich Mendelson; emigrates to Canada to mind his father's fur business, 1921; gains a first in Law, Alberta University,1936; sees action in the Canadian battalion of the International Brigade, 1937-38; becomes trainee manager at Marks and Spencer, Bradford, 1939-41; captured at Tobruk, 1941; organises a mass break-out from his POW camp, 1943; secretary to the Leeds branch of the Communist Party, 1946-53; politicises the Yorkshire miners; a thorn in the side of PM Harold Wilson; on the editorial team of the *World Marxist Review*, 1977-90.

Personal: m. Marion Jessop, author of the feminist history *Petticoat Rebellion*, 1939-67 (her death); m. Joan Smith, 1970-94 (her death).

RICHARDSON, TONY
Born: Jun 5 1928, Shipley, West Yorkshire
Died: Nov 14 1991, California
Director

Born Cecil Antonio Richardson; his film debut is a co-direct with Karel Reisz, *Momma Don't Allow* (1956), a vivid document of trad jazz fever at Wood Green Jazz Club; a leading light of the British New Wave, making *Look Back in Anger* (1959), *The Entertainer* (1960), *Saturday Night and Sunday Morning* (1960), *A Taste of Honey* (1961), *The Loneliness of the Long Distance Runner* (1962), *Tom Jones* (1963), *The Charge of the Light Brigade* (1968), *Ned Kelly* (1970), *The Border* (1982), *The Hotel New Hampshire* (1984), *Blue Sky* (1994); broadly speaking, the good ones are black and white, though *Tom Jones* has its admirers.

Personal: m. Vanessa Redgrave (1962-1967, div.); three daughters – Natasha Richardson (d. Mar 18, 2009), Joely Richardson and Katherine Grimond (her mother is Grizelda Grimond).

SAHNOW, WILL
Born: 1898
Died: Jan 15 1957
General Secretary of the Workers' Music Association, 1939-57

Plays cello and the French horn; gains experience as a conductor with the orchestra of the London Co-operative Society; nurtures brass band music and choral music within the Co-op movement; an early member of the CPGB; a charter member and the first General Secretary of the Workers' Music Association, 1939; organises the Topic Record Club to produce a record a month for WMA members (900-plus at its height); transcribes and arranges a wide range of music for the WMA, including Soviet songs to aid the war effort, community singing songs to boost the Co-op movement, mining songs, Scottish songs, etc; commissions the Keynote series, inc. *The Singing Englishman* by A.L. Lloyd and *Background of the Blues* by lain Lang; instigates and organises the WMA's annual Summer School, from 1949; teaches music in evening classes at Hendon College; dies in his sleep at the outset of WMA's 21st anniversary year, from heart disease exacerbated by overwork, 1957.

SALT, SIR TITUS
Born: September 20 1803, Morley, Yorkshire
Died: December 29 1876, Lightcliffe, Yorkshire
Wool manufacturer, philanthropist, politician

Makes his fortune by developing the market for Alpaca wool; mayor of Bradford, 1848; Liberal MP for Bradford, 1848-50; builds Salt's Mill, 1853; builds the model village, Saltaire, around Salt's Mill; builds Saltaire Congregational Church 1858-59; buried at Saltaire Congregational Church, 1876 – an estimated 100,000 people line the route of his funeral.

Peter Seal (picture by Sheila Seal)

SEAL, PETER
Born: June 15 1959, Aberfeldy, Perthshire
Artist

A Scottish artist long resident in Manchester who works in a variety of mediums – collage, painting and sculpture –as represented by Anthony Hepworth Fine Art; Mike's hand-picked illustrator to *Singing the Century: Bill Leader & Co.*; Mike's joke is that keeping Peter Seal sweet is like keeping James Brown funky.

Personal: m. Sheila; two children, Jack and Rachel, and one granddaughter, Daphne.

Gerry Sharp, Margaret McIver and sons (picture courtesy of Bill Leader)

SHARP, GERRY
Born: ?
Died: Nov 1972
Director of Topic Records, 1957-72

A trained accountant; succeeds Will Sahnow as General Secretary of the WMA, 1957; engineers the separation of Topic Records from the parent organisation, 1960; places Topic on a sound financial footing for the first time in its history, partly by turning the basement of his home in 27 Nassington Road, NW3 into the Topic office and dropping the company's production manager (Bill Leader); consolidates the label's folk direction, relying upon Bill musical judgement and Bert Lloyd's knowledge (Bert, as artistic director of Topic always had the last word), Dick Swettenham's technical wizardry and Jim Boswell's design skills; the 'a' team was topped by Sharp's own financial expertise.

Sally Shuel (picture by Brian Shuel)

SHUEL, SALLY
Born: July 22 1936, London (just off the Old Street roundabout)
Artist, illustrator, businesswoman, copyright expert

Presumably she was born just of the Old Street roundabout because the City of London Maternity Hospital came recommended; it was a long way from Haverstock Hill where the family lived; always known as Sal; the daughter of artist James Boswell; meets future husband Brian at Central School of Arts and Crafts, London, 1954; they marry upon the conclusion of his National Service, Sep 1958; henceforth their personal and professional lives are bound together, jointly designing over a hundred album sleeves for Transatlantic; the iconic drummer boy on the cover of the Watersons' *Frost and Fire* was snapped by Sal; her illustrations are published by *Times Higher Education Supplement*, Age Concern and Gingerbread; manages the estate of James Boswell; the administrator of BAPLA – the British Association of Picture Libraries and Agencies ("I was good at that"); the moving force behind the Collections Picture Library, the photography agency run by the Shuels and son Simon.

SWETTENHAM, RICHARD WYBAULT (DICK)
Born: Jul 21 1927, Bedford, Bedfordshire
Died: Apr 9 2000, Bristol
Audio equipment designer and technical engineer, producer

After National Service, Swettenham studies Physics and Telecommunications Engineering at the London Northern Polytechnic; becomes maintenance engineer and audio equipment designer for EMI, based at Abbey Road, 1951-56; technical trouble-shooter in the EMI team that goes recording opera at La Scala and other opera houses, 1956; chief engineer at Argo, 1956-58; director of Topic Records, 1956 onwards; designs and installs the recording facility at the WMA offices, Bishops Bridge Road, 1956; technical director of Olympic Studios, 1958; designs the world's first professional transistorised mixing console for Olympic's Studio One, 1961; designs a state of the art 12-channel desk console; Rolling Stones, Jimi Hendrix, Dusty Springfield, Troggs, Yardbirds and the Bee Gees beat a trail to Olympic; stealth recording sessions take place for Joe Heaney and Anne Briggs, 1963; redesigns the Olympic studio following relocation to Barnes, 1966; Island Records boss Chris Blackwell backs Swettenham's own company, Helios Electronics, makers of customised consoles, 1969; Island Records is its first client; other customers include the Beatles (Apple Studios), the Who (Rampart), the Rolling Stones (Stones Mobile), 10CC (Strawberry), Richard Branson (The Manor), etc; records Frankie Armstrong, *Lovely On the Water* (12TS216, 1972); authors chapters for the standard training manual *Sound Recording Practice*, 1977; he assists Bill at various technical landmarks, cutting a live disc from an amplified international telephone call from Paul Robeson, 1957, and recording Ramblin' Jack Elliott on board a yacht, Isle of Cowes, 1956; designs the recorder for Topic's first venture into stereo, *The Internationale*, WMA 101, which comes out in mono anyway, 1960.

Personal: m. Eileen (?, separated ?); m. Cynthia (?), an adept in Gardnerian witchcraft; they became High Priest and High Priestess of the Thames Valley Coven.

THISTLETHWAITE, FRANK
Born: May 1 1904, Grassington
Died: ?
Radio engineer and audio innovator

Frank's baptism is solemnised at the Wesleyan Methodist church in Grassington in the West Riding; works as a police photographer, but is more preoccupied with sounds than visuals; supplies Bradford police with a radio communication system of his own design, 1930; builds a recording studio, complete with disc recorder, 1938; hush-hush research work during the war; reestablishes the recording studio in peacetime; persuades audio entrepreneur Gilbert Briggs of Wharfedale Wireless Works to finance the first domestic two-way speaker, 1947 (it takes two men to lift); develops a tape recorder to synchronise sound for 16mm films, circa 1950; Excel Sound Services Ltd opens for business at 49 Bradford Road, Shipley, 1950; Thistlethwaite's showcase deck, the 'Excel', is upgraded and relaunched as the 'Celsonic' with full-track synchronised sound, early c. fifties.
Personal: m. Minetta.

Bill, Bill Snr and José Luis Tovar González

TOVAR GONZÁLEZ, JOSÉ LUIS
Born: San Sebastian, *c.* 1929
A refugee from the Spanish Civil War

One of 4,000 children who arrives at Southampton Docks on May 23 1937 on the cruise-ship *Habana*; fostered by the Leaders, 1940; returns to the care of the Basque Children's Committee upon the move to Keighley, 1940; returns to Spain and either adapts to dictatorship or doesn't.

VAN GYSEGHEM, ANDRÉ
Born: August 18 1906, Eltham, Kent
Died: October 13 1979, London
Theatre director, actor

Works as actor and director at the Embassy Theatre in Swiss Cottage, 1930-34; further experience at the Realistic Theatre in Moscow, 1935; extolls Russia's racial equality to Paul Robeson; directs Robeson in Eugene O'Neill's *All God's Chillun Got Wings*, Embassy Theatre,1933; *Stevedore*, Embassy Theatre, 1935; *A Festival of Music for the People*, Albert Hall, 1939; president of the Unity Theatre; Number Two in the TV series, *The Prisoner*, 1967-68.

Quote: "He was a very considerable theatrical producer, and one of his specialities was being able to fill large halls and deal with the space in a very imaginative way" – Bill Leader.

Selected filmography as actor: *Warning to Wantons*, 1949; *Stop Press Girl*, 1949; *Cromwell*, 1970; *The Pied Piper*, dir. Jacques Demy, w. Donovan, 1972; TV series – *The March of the Peasants*, 1952; *For Whom the Bell Tolls*, 1965; *Sentimental Education*, 1970; *Crown Court*, 1973.

VAUGHAN WILLIAMS, RALPH
Born: Oct 12 1872, Down Ampney, Glos.
Died: Aug 26 1958, London
Composer

Begins to compose at 6; a student of Parry and a friend of Holst at the Royal College of Music; meets his first source singer, Charles Pottipher, 1903; joins the Folk Song Society, 1904; contributes 61 songs to the 1906 issue of the *Folk Song Journal*; editor of *The English Hymnal* – 'Monks Gate' is named after the Sussex village where he collects the tune, 1906; composes *Hugh the Drover*, 1914; edits the *Penguin Book of English Folk Songs* (with A. L. Lloyd), 1959; President of the EFDSS, 1935-1958; the Vaughan Williams Memorial Library, Cecil Sharp House, is named in his honour, 1958.

Selected works: *Toward the Unknown Region* (1905), *The Wasps* (1909), *A Sea Symphony* (1910), *Fantasia on a Theme by Thomas Tallis* (1910), *A London Symphony* (1914), *A Pastoral Symphony* (1922)

Quote: "The opening bars of the *4th Symphony* pin you against the wall" – Bill Leader.

Eric and Audrey Winter (picture by Brian Shuel)

WINTER, AUDREY
Born: Jun 21 1927, Carshalton Beeches, in the London borough of Sutton.
Psychiatric social worker

Places Audrey has done time include West Hampstead, '50-'56, where the girls Jane and Susan wander off and are found by neighbour Johnny Ambrose, and where Audrey declares herself an anti-Stalinist in 1951, to the consternation of the Winters' circle – Eric is branch secretary of the West Hampstead branch of the CPGB; Cambridge, '56-'58, where son Simon is born; Cricklewood '58-'62, where Bill records Archie and Ray Fisher (and Dolina MacLennan and Robin Gray) in the Winters' "nice big room"; Cambridge again at some point; Wandsworth Common – again, we're vague about the precise dates but Audrey can be found managing Neil's Lodge, a restaurant, with daughter Susan between '83-'87, where doubtless her experience running "a free hotel for the folksong world" comes in handy; Mossley, Eric's birthplace, 1987-2004; following Eric's death, returns to London, 2004-present.

Personal: m. Eric 1949; three children, Jane, Susan and Simon.

WINTER, ERIC
Born: Dec 19 1920, Mossley, now Greater Manchester
Died: Oct 23 2000, Mossley
Journalist, librarian, dedicated to peace and socialism through music

Leaves Ashton-under-Lyne Grammar School (a scholarship student) at 15 to be, variously, a trainee manager at Lewis' department store, a wartime RAF recruit, sub-editor on *Home Maker* magazine from the Odhams (later IPC) stable, father of IPC Magazines Group chapels, a librarian at St Pancras Library; branch secretary of the West Hampstead branch of the CPGB; secretary of the UK 'Save the Rosenbergs' committee, 1953; founds *Sing* magazine with John Hasted and Johnny Ambrose, May 1954; writes 'Flowers of Manchester' about the 1958 Munich air disaster and publishes the song anonymously; organises the music for the Aldermaston Marches, 1960-63; campaigns on behalf of his friend, the blacklisted Pete Seeger; folk critic for *Melody Maker* (and, briefly, *New Musical Express*), 1961-1974; refuses to shake Harold Wilson's hand at NUJ annual meeting, circa mid-sixties; teaches journalism at the London College of Printing, 1973-1986; retires to Mossley at wife Audrey's suggestion ("I spent my young manhood trying to get out of the damn place!"), 1987; contributes to *Folk Roots*; produces a songbook in the no-frills *Sing* style of his own songs, *They Never Clapped Me Like That*, 1992; passes away peacefully in his hospital bed with his headphones on, reviewing an album, 2000.

INDEX

The titles of films, books, record albums and classical works are in italics; song titles are in single inverted commas; page numbers in italic indicate a picture of the subject; page numbers in bold indicate an entry in *Possibly Significant People*. The relations of Bill Leader have been given dates to distinguish between all the namesakes. NB Entries in *Possibly Significant People* have not been indexed. It would mean making lists of lists, and there would be no end to it.

39 Steps, The 177
49th Parallel 167-168
'1913 Massacre, The' 33

Abe's Folk Music 123
Accrington, Stanley xxii, *235*, **235**
  declaims in cod-Anglo Saxon, xxvii-xxviii
  consults on a railway technicality, 109-110
Ackroyd, Chris xii, 81, *236*, **236-237**
Adams, Derroll 233
Addison, John 188
Adnopoz, Elliot, see Elliott, Ramblin' Jack
'Affiliate With Me' 67
Airedale Electrical and Manufacturing Co. 105, 108
'Airman's Song, The' 114
Albert Road Junior School 96
Aldermaston March, 151
Alexander, John ('Alex') 7, 89, 224-230, 232, **237**
Alexander, Joyce 89, 226, 229
Alexander, Max 89, 228-231
Almanac Singers, The 32
Almeida, Kirsty xxii
Amalgamated Engineering Union (AEU) 108, 137, 154-155
Amory, Mark 99
Andrews, Dana 166
And So Did I 112
Anthology of American Folk Song, The 213
'Any Old Iron' 48
Anzio, Battle of 89
'Apology for Idlers, An' 190
Appalachia 117
'Apprentice Song, The' 92
Aristocrats (play) 65
Arms for Spain 57
Arthur, Dave xxiv-xxix, xxxiii, 180, *238*, **238-239**
Artists' International Association (AIA) 63
Asch, Moses or Moe, 208-209, 212, **239-240**
Asch, Sholem 208-209
Ashcroft, Peggy 175
Ashley, Clarence 213
Ashpole, Alfred 202

As I Roved Out 124, 206-207
'Asphalter's Song, The' 195
Attenborough, Richard 114
'Ave Maria' 55
Avro Aircraft 84-85

Babes in the Wood (panto) 67
Bach, J.S. 117
'Back Answers' xxvii
Background of the Blues 194
Baker, Anne, née Leader (1880-c.1927) 14, 57
Baker, Dick (b. 1900) 57
Baldwin, Stephen xvii
'Ballad for Americans' 198
'Ballad of the Daily Worker, The' 148
'Ballad of Harry Bridges, The' 32
'Ballad of Harry Pollitt, The' 144
'Ballad of Stalin, The'
'Ballad of Wortley Hall, The' 154, 205
Ballads and Blues (Ewan MacColl's club) 7
Ballads of Sacco and Vanzetti 33
Baltimore Sun, The 37
Banfield, Alistair, 207
Barbirolli, Sir John 119
'Barnyards of Delgatie' 195
Barry, Margaret 218, 233
'Basinful of the Briny' 82
Basque Children of '37 Association 60
Basque Refugees 59-60
Bass, Alfie 66, 169, 189
'Battle of Maldon, The' xxviii
Battleship Potemkin 157, 171-172
Baxter, John 162
Beckton Gas Works 3-4, 8, 15
Beckwith, Reginald 166
Beecham, Sir Thomas xvii
Bedford, The (music hall) 7
Bedford Arms, The (pub) 7
Beeby, Joe *228*, 229
Beethoven, Ludwig 116, 192
Behan, Dominic 233
Bell, Tom 116
Bennett, Alan (French teacher) 127
Bergan, Ronald 172
Bergman, Ingmar 183
Betbeder, Camilla, see Raab, Camilla
Betjeman, John 63
Betts, Pete xxxiv
Bert: The Life and Times of A.L. Lloyd xxv, 180
Bexley Square, Battle of 162
Bigg, A.E. 4
'Bill Dalton's Wife' xxxiii
Birdman 177
Birds, The 177
Birmingham Clarion Singers, the 63
Bishop, Peter 142
Bishop's Candlestick, The (play) 174
Bishops Bridge Road (17), W2, 153
Black April 30
Black, Cilla xxix
Black Horse Broadside, the 160
Blackball Line, The 217-218
Blackletter Ballads 123-124
Blackshirts 60
Blair, Tony xxvii
Blavatsky, Madame 27
'Bless 'Em All' 68
Blue Lamp, The 171
Blue Note Records 224
Blumlein, Alan 175
Blunt, Anthony 187
Bogarde, Dirk 171
Boggs, 'Dock' 213
'Boiled Beef and Carrots' 48
Bolton, Felicity 202, *204*, 205
'Bonny Boy, The' 197
Bootle, David xxviii
Boston (book) 37
Boston Traveller 35
Boswell, James (Jim) 60, *61*, *62*, 63, 87, 169-171, 232-233, *240*, **241**
Bothy Band xxxiii
Boulting, John 189
Boulting ,Roy 187, 189
Bowie, David 166
Boyd, Joe xvi
Brecht, Bertold 64

Bradford Civic Theatre 116
Bradford College of Art 122
Bradford Royal Infirmary 101
Bradford Technical College 105
Brandenburg Concerto no. 2 117
Brando, Marlon 137, 167
Bridges, Harry 32, **242**
Briggs, Anne xiv
Briggs, Gilbert 103
Briggs, Norman 161
Brighton Rock 189
Brinson, Peter 181-182, 184, **242-243**
Britain and the Spanish Civil War 57
British Council 181
British Road to Bolshevism, The 7
Britten, Benjamin xxiv, 195
Brocken, Michael 219
Brontē, Branwell 109-110
Brook, Peter 176
Brooke, Rupert 52
Brooks, Ern 232
Brown, William 70, 73
'Browned Off' 88
Brownie McGhee and Sonny Terry 210
Bryan, Dora 189
Bryce, Owen *203*, 205, **243**
Buchan, Janey 196
Buchanan, Tom 57
Bulletin, The 64, 191-192, 196, 199-201-203, 205
Bunkum Entertainments 227
Bunuel, Luis xxiv, 183
Burgess, Guy 187
Bush, Alan xxiv-xxv, 64-65, 68-69, 129, 133, 142, 186-188, 197-199, 202, *204*, *243*, **244**
Bush, Nancy 64
Butler, Mike xii, xv-xviii, xxx, xxxiv, 129, 225, **245**
Butler, Rab 99, 101
Butlin's Holiday Camp, Filey 108, *121*, 122

Caedmon (label) 175
Cabinet of Dr Caligari, The 166
Cable Street, Battle of 60, *61*
Calcott Walk (1), Mottingham, 70, 74, 76
Cameron, Isla 207, 233
'Camden Girls' 169, *170*
Campbell, Ian xxix, 92
'Canadee-i-o' xxx
Canterbury Tale, A 165, 167
'Captain Wedderburn's Courtship' xxxiii
Carlton Grammar School 127, 252
Carry On Henry 108
Carthy, Eliza xxii
Caxton Publishing Company 110
Cecil Sharp House xxi-xxii, 173
Chamberlain, Neville 67, 69
'Champagne Charlie' 7
Chant du Monde (label) 218
'Chapayev' 194
Charters, Sam 209-210
Childhood of Maxim Gorky, The 158
Child in the House, A 167
Chopin, Frédéric 116
Church, Esme 116
Churchill, Winston 143, 179
Citizen Kane 161, 174
City Ramblers, the 201
Clancy, Willie 218
Clooney, Rosemary 198
Closely Observed Trains 185
Coaldust Ballads 195
Coe, Pete 123
Coen brothers (Joel and Ethan) 30
Coleman, Elizabeth Brooke 31-33, 37, 39
Coleman, Elizabeth (Betty) *36*, 37
Coleman, Michael (fiddler) 39
Coleman, Nancy *36*
Coleman, Robert Henry 31, *36*, 37
Coleman, Susan, see Mooney, Mrs Richard E.
Coleman, William Caldwell (Judge) xviii, 30-33, 35, *36*, 37-39, **246**
Coleman, William Caldwell (son) 31, *34*, *36*, 37
Coleman, William Caldwell III *36*
Collins, Judy xxix
Collins, Shirley xxii, xxxiii, 122

Columbia label (UK) 223
Columbia label (US) 215
Colvin, Claudette 179
Come All Ye Bold Miners 9, 123-124, 195
Come Along John 222
'Come on-a My House' 198
Common Touch, The xviii, 162-166
Communist Manifesto, The (a co-write with Friedrich Engels) 27
Communist Party of Great Britain (CPGB), 43, 57, 120, 140, 142-146, 180, 186-188,
music group of, 196-197
Congress of Industrial Organizations, The 32
Connolly, Billy xiv
Connolly, Ríoghnach xiv, xxii, xxix
Conquest of Everest, The 171
Conway Hall 27, 141
Coolidge, Calvin 32
Co-operative Society 67, 151, 154, 187, 195, 206
Corum, Alf 196, **246-247**
Corum, Wendy *204*, 247
Costello, Cecilia xvii
Cotten, Joseph 176
'Country Gardens' 73
Coward, Noel 114, 267
Cox, George (1907-1962) 23, 45, 48
Cox, George (1931-1997) 23, 76
Cox, Harry 207
Cox, Lilian (Lily, née Leader) (1906-1973) *13*, 19, 44-45, 76
Cox, Lilian (Lily) see Niles, Lilian (Lily)
Crane, Walter 223
Cranford 5
Crompton, Richmal 70
Cronshaw, Andrew xxx
Cutler, Ivor xx

Daily Mirror, the 156
Daily Worker, the xviii, 94, 148, 197
Dallas, Gloria (née Whittington, née Leader) xxvi-xxvii, 21, 25, 32, 38-39, 70, 111, 113-114, *115*, 116, 118-120, *120*, *121*, 122-125, 127, 129, *132*, 161, 172, 177, 180, *204*, 207, 210, 220, *247*, **247-248**
Dallas, Karl xxvi-xxvii, xxx, xxxiv, 86, 141, *248*. **248-249**
Dallas, Stephen 86
Daily Worker, The 56-57, 63, 84-85, 119
Dance and Dancers 184
Danú xxxiii
Daphnis and Chloe 118
Das Kapital 27, 140
Davenports Magic Shop 230-231
Davies, Sean 174-177, 222-223, **249**
Davis, Anne (Nance, née Leader) (1892-c.1975) 11, *13*, 19, 21-25, 28, 38-39, **250**
Davis, Cathy (b.1949) 38
Davis, Harry (1893-?) *13*, 23, 38-39
Day, Doris 198
Daybreak 67-168
Days of Contempt 63
'D-Day Dodgers' 88
'Dead, The' 225
Dead of Night 173
Dean, James 180
Debussy, Claude 118
Decca Records 220
Decibel (label) 170
De Dannan xxxiii
Deep Lancashire xxxiii
Delius, Frederick 117
Dickens, Charles 9, 94
Dickson, Barbara xxxi
Did You Kiss the Foot That Kicked You 141-142
Discreet Charm of the Bourgeoisie, The xxiv
Disney, Walt 76
Disraeli, Benjamin 94
'Don't Dilly Dally On the Way (The Cock Linnet Song)' 81, 114
Dot and Carry One 84, 153
D'Oyly Carte Opera Company 23
Driberg, Tom 108-109, 188, *250*, **250-251**
'Drumdelgie' 123

Dubjax 163
Dubliners, The (book) 225
du Maurier, Daphne 176
Dunman, Jack 187
Dunn, George xvii
Durbin, Deanna 160
Durkin, Tom 107-108
Dyke, Greg xxiv

Ealing Studios 159, 169-171
Eastbrook Hall 116
Eaton, Alex xxx, xxxii, 84-85, *115*, 126-129, *130-132*, 133-134, 151, 153-154, *204*, 217, **251-252**
Eaton, Felix 154-156, **252**
Eaton, Louise, *115*, 126-129, *131-132*, 133-134, 138, 154-156, *204*, *251*, **252**
Eclipse (label) 221
Economist, The 233
Eden, Anthony 179
Education Act, 1944, 97, 99-100
'Ee When I Were a Lad' xxxiv
Einstein, Albert xv, 179
Eisenhower, Dwight D. 198
Eisenstein, Sergei 157, 172, **252-253**
El Alamein, Battle of 181
'Election Stamp, The' 99
Elen, Gus 8
Elgar, Edward 117
Eliot, George 182
Elizabeth II 126-127
Ellington, Duke 197
Elliott, Ramblin' Jack 33, 217, 228, 233
Ellis, John xiv-xv, xxxii, 153, **253**
Ellis, Ruth 179
Elmstead Wood Junior School 74
Elysian Jazz Band, the 217
Emmerdale Farm 106
Empire Theatre of Varieties, The (music hall) 7
Engels, Friedrich 27, 116
Engle, Tony xxx, 6
English Consort Orchestra, 133
English Dance and Song (magazine) xxvi
English Electric 108, 138
English Folk Dance Society xxii-xxiii, xxv
English Folk Dance and Song Society, see also Cecil Sharp House xxi-xxii, xxxii-xxxiii
English Folk Revival, The 219
English Songs Vol. 2 122
Enigma of Nic Jones, The xxvi
Enigma Variations 117
Esholt Waste Water 106-107
Eureka Brass Band 210
Evans, Bill 208
Ewan, Ruth 141
Excel Sound Services Ltd 103
Eye, The 60

Falcon, Cleoma 213
False True Lovers 122
Falstaff 117
'Family of Man, The' xxvi
Fantasia 160
'Feast of the Mau Mau, The' 148-149
Fellini, Federico 183
Festival of Music for the People, A 67
Few Tunes of Good Music, A 5-6, 170
Fields, Gracie 117, 159, 221, **253-254**
Film (Manvell) 164
Film Criticism and Caricatures 1943-53 158
Films and Filming 184, 214
Films of Poland xvi, 172-173, 180-186, 191, 199
Finnegans Wake (book) xvi
Firestein, Jack 60, 89, **254**
Fisk Jubilee Singers, the 210-211
Fitton, James 63, 169
Flanagan and Allen 77
Fleming, Ann 99
Folk-Song Society xxii
Folksound of Britain 222
Folkways Records 186, 206-207, 209-213
Ford Dagenham 44, 55, 139
Foreman, John 7-8, 66, 86, 88, 119, 217, 226-228, 231, **255-256**
'Four Loom Weaver, The' 195

'Fourpence a Day' 195
Francis, Harry 197
Franco, Francisco 53, 59
Frame, Pete 87
'Freight Train' 201
Frenzy 177
Friel, Desi xxix
Friel, James 63, 87
From the Hebrides: Further Gleanings of Tale and Song xxii
Frow, Edmund 162
Foyle's xxix

Gaimon, Bridget (1844-1916) 2, 11, *12*, 14, **256**
Galvin, Patrick 195, 197, **256**
Game of Hide and Seek, A 178
Gansler, James xxix, xxxiii
Gansler, Jay xxix
Garbutt, Vin xxxiv
Gaskell, Elizabeth 5
Gaslight 168
Gates, Bill 179
Gates, Rev. 213
Gentle Gunman, The 171
Geordie 173
Gielgud, Sir John 175-177
W.S. Gilbert 23
Glasgow University, Political Song Collection, 64, 196, 201, 211
Glass, Philip xv
Gold Badge award xxi-xxii, xxxii-xxxiii
Gold Folk Show, the xxiii-xxiv, xxxiii
Golden Age of Road Travel, The (TV) 82
Goldsbrough, Arnold *132*, 133, 194, 202, *204*, **257**
Goldstein, Kenneth S. 207-208, **258**
Gollancz, Victor 63
Gondoliers, The 23
'Goodnight Irene' 199
Goon Show, The 181
Gorman, Michael 218, 233
Goss, John 205
Gould, Diana xxvii
Grainger, Percy xvii, 73
Granceta 107
Grand Prix of Gibraltar, The 208
Graney, Paul, 87, 162, **258**
Grant, Cary xxiii
Grant, Cy 88-89, *259*, **260**
Grant, Ulysses S. 210
Graziani, Marshal 88
Great Palestinian Revolt, the 59
Great Zadora, The 227
Green, Craftsman *125*, 126
Green, Tony 123, **260**
Greenwood, Walter 161
Grossman, Stefan 122
Griffith, Kenneth 187
Guiseley Station 109
Guthrie, Woody 32-33, 35, 124
'Guy is a Guy, A' 198

Habana SS 59
'Ha Ha, This-A-Way' 198-199
'Hal-an-Tow' xxxiii
Haley, Bill 179
Haley, Kate 122-123
Hall, Martin xxvi, **261**
Hall, Reg 5-6, 153, 167, 170, 206-207, **261-262**
Hallé Orchestra 119, 202
Halliard, the xxvi
Hambourg, Mark 163-164
Hamer, Robert 169
Hancock, Tony 214
Hand in Hand 198
Handel, George Frideric 55
Happy Family, The 173
Happy is the Bride 189
Harbour, Amy (b. 1977) xxiv
Hardcastle Crags 113-114, 116, 128-129
Harding, Mike xxii, xxiv, xxxi-xxxiii, *262*, **262-263**
Hargreaves, John 194
Harris, Herbert 207
Harris, Mr 4
Harrison, Kathleen 173
Harte, Frank 133
'Has Anyone Seen Grandad?' xxix

Hasted, John 99, 123, 129, *132*, 138, 144, 148, 154, 185-186, 202, *203*, *263*, **264-265**
Hawkins, Screamin' Jay 149
Hays, Lee 32
Hecht, Philip 202, *204*
Help Spain: Voluntaries, Británicos e Irlandeses en la Guerra Civil Española 57
Henderson, Hamish 88, **265-266**
Hennessy of Nympsfield, Baron 141-142
Hepton, Bernard 116
Hibbert, Geoffrey 162-163, **266-267**
High Treason 187-189
Hill & Dale (label) xv
History of Mr Polly, The 173
Hitchcock, Alfred 176-177, **267**
Hitler, Adolf 33, 53, 59, 67, 144, 163
Hobsbawm, Eric 141, 219-220, **267**
Hobsbawm Files, The (radio) 141
Hockney, David 122
Holdridge, Barbara 175
Holiday, Billie 141
Holland, James 63
Holloway, Stanley 173
Holtby, Winifred 25
Homophone 220-221
Homage to Catalonia 57
Hornby P.C. 66
Horning 45, 77
Horowitz, Vladimir 164
How to Make Your Husband a Sultan 207
Howard, Leslie 168
Howarth, John xxvii, xxix, xxxiv, **268**
Howells, Kim xvii
Hughes, Langston 30
Hugh the Drover (opera) xxiv, 119
Hull, Will 228
Hunter Muskett xxx
Hyett, Trevor xxiii

IBC (International Broadcasting Company) 174-175, 222
I, Claudius (audiobook) 177
I Confess 177
If I Remember Writely: The Unsound Memories of a Soundman 14, 21, 33, 43-44, 56, 59, 78, 83, 96, 99, 102, 170, 179
'If A Grey Haired Lady Says "How's Your Father"' 77-78
'If It Wasn't for the 'Ouses in Between' 8
I-Hsuan, Yu 198
'I Just Want to Sing Your Name' 33
'I Love My Miner Lad' aka 'Miner Lad' 122-123
I'm Alright, Jack 189
'I'm Champion at Keeping Them Rolling' 195
Importance of Being Earnest, The 173
Iñárritu, Alejandro G. 177
'In Spite of Ourselves' xxix
International Folk Song Contest 218
In the Fight For Spain 218
Introduction to Gospel Song, An 210
In Which We Serve 114, 162
Irish Songs and Dances 217
Irish Songs of Resistance (book) 195, 197
Irish Songs of Resistance (records; two volumes) 218
Irvine, Alexander xxxvi, 25
Islamic Liturgy: Song & Dance at a Meeting of Dervishes 212
It Always Rains on Sunday 165, 169-171
It's a Wonderful Life 183
It Was Mighty 170

Jackson, Peter 97-98
Jacobi, Derek 177
Jamaican Cult Music 212
James, Henry 30
Jansch, Bert xiv-xv, xxix
Jazz Scene, The 219
Jeffersons, the 221
Jerome, Jerome K. 166
'Jerusalem' 81
Jesus Christ 24, 27, 96
Jeux Interdits (*Forbidden Games*) 183
John, Elton 169
Johns, Glynis 168

Johnson, Blind Willie 213
Jones, Jim (1912-1980) 45
Jones, Julia xxx
Jones, Kathleen (Kit, née Leader) (1914-1972) 19, 24, 45, *50*, 76, 78, 80
Jones, Mary née White (b. 1936) 24
Jones, Nic xiv, xxi-xxii, xxvi, xxx, *268*, **269**
Jones, Peter (actor) 189
Jones, Philly Joe 208
Jordan, John 196
Joseph, Nat xvi
Joyce, James 225
Judd, Bartlett 22, 26
Judd, Harold 27
Judd, Mary 26-28, **269**
Jungle, The 30-31

Karpeles, Maud xxii
Keene, Steve xxvi
Kelly, Stan xvi, xxiv-xxvi, xxx, xxxiii-xxix, 117-119, *270*, **270-271**
Kenna, Andy xxviii-xxix, **271-272**
Kennedy, Douglas xxii
Kennedy, Helen xxii
Kennedy, John F. 160
Kennedy, Peter xxii, 124, 207
Kennedy-Fraser, Margaret xxii
Kerr, Deborah 162
Khrushchev, Nikita 145, 180
Kidson, Ethel 123
Kidson, Frank 123
Killoran, Paddy 39
Kimber, William xxiii
Kind Hearts and Coronets 173
King George V 54
King George VI 54, 126
King Street (16), WC2, 142, 144
Kingston Trio 199
Kingsway Hall 55
Kinsey, Ray 55, 90, **272**
Kisch, Royalton 188
'Knave is a Knave, A' 198
Knight Without Armour 173
Kydd, Sam 189

Labour League of Youth 108
Lambert, Richard S. 96
Lampell, Millard 32
'Land of Freedom, The' 65
Lang, Iain 194
Lassalle, Ferdinand 27
Last Battle, The (book) 166
Last of the Summer Wine 67, 166
L'Atalante 167
Lawrence, Martin 198
Lawrence and Wishart 60
Leader, Agnes, née Woodcock (1869-1950) 11, *14*, 75
Leader, Amy, see Harbour, Amy
Leader, Anne, see Baker, Anne
Leader, Anne (Nance), see Davis, Anne (Nance)
Leader, Annie (b. 1993) xviii
LEADER, BILL *40*, *41*, *50*, *58*, *71-72*, *100*, *115*, *124*, *131*, *132*, *139*, *147*, *204*
  1955, his hot year, 172
  and actors, 172-173, 176-177
  on John Alexander, 224, 227-228
  and ancestors, 1-2, 6
  anti-royalism, 54
  apprenticeship, 104-105, 107, 109, 122, 148
  atheism, 54
  on aunts and uncles, 45-49
  in Becontree, 55-56
  birth and childhood, xxiii, 42
  and Joe Boyd, xvi
  on Bradford's cultural life, 116, 118-119
  Bulletin, 199-200, 205
  at Calcott Walk, Mottingham, 70, 74, 76
  on Canning Town, 9
  childhood reading, 73
  his cinema habit, 159-169
  and civil defence, 75-76
  colour-blindness, 109
  on communal singing, 81-82
  and the CPGB, 142, 186-188
  courtship and marriage to Gloria,

111, 114, 120, 122, 124-125
and *the Daily Worker*, 148
on death, xix
early trades, 104-105, 108, 110
and Alex Eaton 128-129
education 71, 73-74, 96-97, 101-102
on empire and etymology, 54, 82
on ethnicity, 32
as Everyman, xiii
on family relations, 1, 3, 76
fear of shadows, 73
attends *A Festival of Music for the People*, 67-68
Films and Filming 184
and Films of Poland, 172-173, 179-186
on Folkways, 208-213
frustrated audio ambitions, 102
and the Gold Badge/Gold Folk Show, xxi-xxiv
health, xxiii, 101
in Hong Kong, 108, 124, 126, 149-151
in Keighley, 59, 80-83
and Ray Kinsey, 55,
a life in bullet points, xxxii-xxxiii
moves to London, 134
on Lou: her legend and her reading, 25, 27, 28, 30-31, 86
and the May Day march, 137
meeting Gloria 114, 116, John Alexander 224-226, MacColl 234, Leon Rosselson 185, Eric Winter 185
music education and skill, 116-118
National Service 124-127, 133, 148-151, *152*
nicked for writing graffiti, 145-146
"Nothing truthful has ever been written..." 29
political education, xviii, 56
public speaking, 148
on radio and television xxvi, xxxiv
on Sacco and Vanzetti, 35
and Will Sahnow 192
his scepticism, 39, 49, 80
sex education, 105-106
as a songwriter 154-156
on state surveillance, 139-140
on 'stretching', 171
on stripeys and his surprise ignorance thereof, 233
as a talent spotter xiv
Thwaites Brow epiphany, xii, 82-83
and Topic Records, xiv, xvii, 199
on Topic economics, 219-221, 223-224, 231-232
and José Luis Tovar González, 59
on Unity Theatre, 65-67
and VAT, xiii-xiv
and the Winters, xxix-xxx
wit, wisdom and musical judgement, 33, 197-199, 205, 207-208
and the WMA, xvii, 129, 151, 153, 157, 191, 196, 216
at the WMA Summer School, 1958, 156, 185, 202
on the parallels between the WMA and Elgin Music Society, 187-188
and the Young Communist League, 113
albums, EPs, CDs (as engineer, producer, selector)
Border Minstrel 133
Bright Phoebus 222
Cloud Valley xxx
Dublin Street Songs 133
Grand Airs of Connemara, The xvi
Harmonica Blues 210
Lancashire Lad, A xxxiii
Men at Work: Topic Sampler No. 3 33
Music of New Orleans Volume 1, The (Topic) 209-210
Music of New Orleans Volume 3, The (Topic) 209-210
Outback Ballads 233
Pete Seeger's Guitar Guide for Folk Singers Play 211

Pete Seeger's Play the 5 String Banjo 211
Prosperous xxxiii
Rambling Boys, The 233
Round and Round with the Jeffersons 221
Second Shift 233
Spirituals (Fisk Jubilee Singers) 210
Still I Love Him 233
Street Songs and Fiddle Tunes of Ireland 218, 233
Streets of Song 233
Takes the Floor 233
Unto Brigg Fair xvii
Woody Guthrie's Blues 217-218
Leader, Bill, *father* (1888-1961), xviii, 1-3, 14, 16-17, 38-39, *40*, 42-43, 48, 57, *58*, 67-68, 74, 83, *84*, 98, *100*, 101, 104, *115*, 117, *124*, 125, 139-140, 160, **278**
Leader, Bridget, née Ryan, see Gaimon, Bridget
Leader, Bridget, see Waldron, Bridget
Leader, Daniel (Danny) (1911-1993) *50*
Leader, Edward James (Ted) (1908-1983) 19, 47-48, **272-273**
Leader, Eleanor Rose (Elsie), née Holtum (1909-1987) 79
Leader, Grace (?-?) 48-49
Leader, Henry (Harry) (1900-1908) 16, 19, 46
Leader, Helen, née Heron, xxx, xxxii, 109, 153, **273**
Leader, Henrietta (Hetty) (1884-1973) 48-49
Leader, James (c.1884-1918) 14, 16
Leader, James William (Jim) (1911-1980) 19, 23-24, 28, 46-47, 79
Leader, Jane or Joanna, née McNamarra (1851-1922) 2, 14-15, 17
Leader, Jo, see Pye, Jo
Leader, John (Jack) (1884-1950) 14, 48
Leader, Julia, née Hurley (1871-1953) *50*
Leader, Catherine (Kate) (c.1893-1938) 14, 16
Leader, Kathleen (Kit), see Jones, Kathleen
Leader, Lilian, see Cox, Lilian
Leader, Louisa Jane (Lou) (1896-1974) xviii, 1-3, 11, 19, 21-22, *34*, 39, *40*, 44-49, 55-57, 67-68, 80, 83-86, 94, 98, 101, 104, 117, 122, *124*, 125, 140, 148, 160, 166, *274*, **274**
and the Colemans 31-32
her reading tastes, 25-26, 28, 37
in the USA, 28-43
Leader, Louisa Martha, née Parsons (1871-1954) 2, 9, 19, 23, 25-26, 45, 75-76, 79-80
Leader, Lynne, née Porter, xxiii-xxiv, 45, **275**
Leader, Margaret (Maggie), née Wicks (1903-1992) 11, 15, 23, 48, 59, 75, 78, 140, 160, **275-276**
Leader, Mary Elizabeth (Molly), see White, Mary Elizabeth (Molly)
Leader, Michael (1892-1916) 14, 16-17
Leader, Michael (b.1946) 79
Leader, Patricia, see Smee, Patricia
Leader, Patrick (1839-1877) xviii, 1-3, 6-7, 11, 14, **276**
Leader, Patrick (1869-1933) 2, 5, 11
Leader, Patrick (1881-1962) 5, 14, 16
Leader, Patrick (1903-1980) 5, 19, 24, 46, 48, 75, 160, **276-277**
Leader, Patrick (1890-?) 5
Leader, Sally (1906-2001) 2, 15
Leader, Thomas (Irish patriarch) (c. early 19th century) 1
Leader, Thomas (c. 1847-c. 1902) xviii, 1, 3, 6-7, 14-15, **277**
Leader, Thomas (1865-1942) 2, 11, *13*, 15-16, 23, 25-26, 75-77, 80, **277**
Leader, Thomas (1877-1953) 14, *50*
Leader, Thomas (1894-95) 19
Leader, Thomas, Tommy (1905-c.1948) 48-49, *50*
Leader, Tom (b. 1959) xviii, xxxiii, 25, 46, 141
Leader, Violet, née Ingrey (1916-2010) 23, 47-48

Leader, Walter Thomas (1891-1891) 19
Leader (label) xiii, xvii, xxxii, 184, 213
Leader Sound (studio, Greetland) xxxii
Leader Sound (studio, 5 North Villas) xv-xvi
Ledbetter, Huddle (Leadbelly) 199
Lee, Jack *278*, **278**
Leech, Eric (Youngfellow) xii
Left Book Club, The 63
Left Review, The 63
Lehmann, Beatrix 166, 174
Lenin, Vladimir 7
'Let the Bright Seraphim' 55
'Let's Walk Tha-A-Way' 198
Leviathan, the (ship) 28-29
Levin, Ariel 56, **279**
Levin, Deana **279**
Lewis, C.S. 166
Lewis, Sinclair 31, 37
Lief, Craftsman *125*, 126
Life and Death of Colonel Blimp, The 168
Limeliters, The 144
'Limerick Rake, The' 154
Little, George 87
Little, Marie 86-87
Littlewood, Joan xxv, 109
'Liverpool Lullaby' xxviii-xxix, 117-118
Livingston Studios 55
Livingstone, David 94
Lloyd, A.L. (Bert) xxiv-xxv, 114, 123-124, 140-141, 180, 194-195, 207, 217, 233
Lloyd, Joe xxviii
Lolita 180
London Labour and the London Poor 3
London Symphony Orchestra 118
London Youth Choir 185-186, 198
Long Memory, The 165, 167
Lough, Master Ernest 54-55
Lovell, Raymond 163
Love on the Dole (film) 161-162
'Love on the Dole' (song) 67
Lumière brothers, Louis and Auguste, 171
Lutyens, Edwin 16
Lynott, Martin xxvi, *280*, **280-281**
Lytton, Henry 23

MacColl, Ewan xxv, 7, 87-88, 114, 124, 140-141, 145, 162, 195, 207-208, 217, 233-234
MacGinnis, Niall 165-166, 168
Maclashan, Lance Corporal *125*, 126
Maclean, Donald 187
Magic Box, The 173
Magnegraph 208
Maguire, Donal 108
Mahler, Gustav 118
Making of the English Working Class, The 140
Malarky, Father 24
Malaya Emergency 149
Maldon Folk Club xxviii
Malleson, Miles 173-174
Malraux, André 63
Manchester Guardian 143
Man In the White Suit, The 173
Manvell, Roger 164
Margaret, Princess xxii
Marks and Spencer's 85
Marshall, Herbert 66
Martin, George xv
Marvels Lane School 74
Marx, Karl 27, 93, 140, **281**
Marzials, Theo 22
Massey, Raymond 168
Mayhew, Christopher 179
Mayhew, Henry 3
McCallum, John 169
McDevitt, Chas 201
McDonald, Jacqui 123
McElhiney, Kirk xxix
McGhee, Brownie 210, 212
'McKafferty' 195
McShee, Jacqui xxvi
Meisel, Edmund 172
Melodisc Records 209
Melody Maker 186
Menuhin, Yehudi 80
Met, The (Bury), xxiii

Metropolis 172
Metropolitan, The (music hall) 7, 224
Meyerhold, Vsevolod 66
MI5 140-141, 144, 187
Mickey Mouse 76, 160
Middlemarch 182
Midsummer Night's Dream, A 114
Mikado, The 23
Miles, Bernard 163
Miller, Jimmie, see MacColl, Ewan
Mills, John 114, 165, 167, 173
Minchinton, John 181-184, **281-282**
Miners' Strike, 1984-85, 94
Miserables, Les 174
Mitchell, Guy 199
Mitchell, Miss 119
Moisenco, Rena 194
Molière 174
Monk, Thelonious 106, 208
'Mon Like Thee, A' xxix
Monteagle Infants School *71*, 73
Monte Cassino, Battle of 46
Monteux, Pierre 118, **282**
Mooney, Mrs Richard E., née Coleman, Susan 31, *36*, 37
Moonlight Sonata 220
Moore, Christy xiv, xxxiii
More R& B From S & B 212
More Songs of the Hebrides xxii
Moreton, Ronnie 106, *107*, 108, 138-139, *139*
Morning Star, the see *Daily Worker*
Morrison, Herbert 85
Moscow Trials 53
Mosley, Oswald 54, 60
Moss Bros 186
Moss, Henry xxiv, **282**
'Mrs McGrath' 197-198
Mushroom Ceremony of the Mazatec Indians of Mexico 207, 209
Music and Life 196-197, 199
Music of New Orleans Volume 2, The (Melodisc) 209-210
Music and Society 194
Musicians' Union 197, 219
Muspratt-Dunman, Helen 187
Mussolini, Benito 53, 67
'My Bonny Miner Lad' 122
My Lady of the Chimney Corner xxxvi, 25-26

Nash, Paul 187
Nassington Road (27), NW3, 141, 233
National Service 108, 124-127, 133
National Unemployed Workers' Movement 162, 172
National Union of Agricultural Workers 187
Navarro López, Eva Maria *283*, **283**
Nazarin 183
Neat, Timothy 88
Negro Folk Music of Africa and America 212
New Era Film Society 157, 172, 180
New Symphony Orchestra 188
Newton, Francis, see Hobsbawm, Eric
Newton, Frankie 141
New Victoria, Bradford 161
Next Time Round 123
Nicholls, Anthony 188
Nichols, Dandy 173
Night of the Demon 165-166
Niles, Lilian (Lily) (b.1935) 76, 79
Noakes, Rab xxxiv
North by North West 177
North Villas (5), NW1, xv

Oakmar, SS 32
Oddfellows Arms, The xv, xxi-xxii, xxiv, xxvi-xxvii, xxxiii
O'Donnell, Bridie 123
'O For a Wings of a Dove' 54
Off to Sea Again 217-218
Ogden, Ron 172, 179-180
'Oh, For a Closer Walk With God' 55
Oil 30
Okhlopkov, Nikolay 66
'Old Brown Sat in the Rose and Crown' 77
'Old Man's Song, The' xxix

Olivier, Laurence 168, 175, 198
O'Neill, Eugene 161
Once a Jolly Swagman 168
'On Top of No Smoking' 99
Orwell, George 57, 63
Owen, Bill 67, 168, 171, **283-284**

Pack, Charles Lloyd 187
Pageant of Co-operation, A 68
Pageant of Labour, The 68
Palace Theatre (music hall) 7
Panufnik, Andrzej 181
Park Grove (13), Shipley, 96, 118, 140, 276
Parker, Charlie 106
Parker, Evan 129
Park Grove, Shipley, 96
Parks, Rosa 179
Parsons, Walter 19
'Parting Glass, The' xxxiv
Pasezerka 183
Passing of the Third Floor Back, The 166
Passport to Pimlico 165
Peace Council, the 56, 139
Pearl Harbour 33
Pears, Peter 195
Pegg, Bob 123
Penguin Eggs xxx
Penguin Film Review 165
People 142
Persephone (dance company) 127
Peterkin, Julia 30
Peterloo 162
Piano Concerto (Tchaikovsky) 163-164
Pigg, Billy 133
Pilgrim's Progress (opera) 118
Pinocchio 160
Pinter, Harold xviii
Pioneer Songbook, The 195
Piper, John 169
Pity and Terror: Picasso's Path to Guernica 57
Plant in the Sun 66
Planxty xxxiii
Plays and Players 184
Plessey workers' anthem 192
Poe, Edgar Allan 70
Pogodin, Nikolai 66
Poland Today 184
'Policeman's Holiday, The' 69
Polish Cultural Institute 180-181, 184
Politkovskaya, Anna xxvi
Pollack, Anna 197
Pollitt, Harry 57, 144-145, **284**
Pool of London 171
Popular Soviet Songs 64, 195
Porridge 166
Portman, Eric 167-168, **284**
Potts, Darren xxiv
Powell, Michael 167-168
Powles, John 196
Prague Spring, 1968, 197
Prague Spring of International Festival of Music, 1959, 197
Prague Theatre of Music, 197
Pressburger, Emeric 167-168
Press Gang, The, or *The Escap'd Apprentice* (opera) 64
'Pride of the Coombe, The' 108
Priestley, J.B. 63
Princess Alice 4
Princess Margaret 180, 207
Prine, John xxix
Prisoner, The (TV) 67
Private Life of Henry VIII, The 173
Prohibition 32
Prosky, Irving 207
Proudfoot, Michael xxvi
Proudhon, Pierre-Joseph 27
Psycho 177
Pudovkin, Vsevolod 157
Purchase Tax, 216, 223-224
Pybus, Mr 117
Pye, Jo, née Leader (b.1942) 5-6, 9, 11-12, 14-17, 21-26, 28-29, 37-38, 44, 46-49, 59, 75, 78-80, 122, 137-138, 140, 148, 160-161, 187, *285*, **285**
Pye, Mike 5-6

'Quartermaster's Store, The' 68-69, 114
Quaye, Russell, 87, *286*, **286**

Queen of Hearts 159
Queen of Spades, The 173
Queen Victoria 210

Raab, Camilla 153, 189, 191, 199, 234, **287**
Rafferty, Gerry xiv
Rainbow Jacket, The 171
Ramelson, Bert 85, **287**
Ravel, Maurice 117-118
Ray, Johnnie 198
Ray, Man 187
RCA Victor 215
Realistic Theatre 66
Reed, Blind Alfred 166
Reed, Maxwell 167
Reed, Oliver 167
'Rejoice Greatly' 55
Renbourn, John xv, xxvi
Restless Generation, The 87
Rex Records 221
Reynolds, Ian xxvi, xxix, xxxiii
Richardson, Clarence Albert 98-99
Richardson, Ralph 175
Richardson, Tony 98-99, **288**
Rigby, Edward 163
Rin Tin Tin 159
'Rising of the Moon, The' xxix
Rite of Spring, The 118, 160
Riverside Records 206, 208
Road to Wigan Pier, The 63
Robertson, Jeannie xvii
Robeson, Paul 66, 68-69, 217
Robinson, Earl 198
'Rochdale Cowboy, The' xxxiii
Rogge, Michael 149
Rogue Male (book) 168
'Roll in My Sweet Baby's Arms' xxxiii
Rollins Sonny 208
Roney, Marianne 175
Rope 177
Rosenberg, Ethel 143
Rosenberg, Julius 143
Rosselson, Leon 185
Rothschild, Lord 175
Rothshild, Paul xvi
'Roving Kind, The' 199
Row, Bullies, Row 217-218
Rowbotham, Bill, see Owen, Bill
Royal Academy of Music, the 64
Royal Albert Hall 67-69
Royal Electrical and Mechanical Engineers 108, 149
Ruskin, John 94
Rutherford, Margaret 173

Sacco, Nicola xviii, 33, 35, 37
Sadlers Wells 197
'Sailor Cut Down in His Prime, A' 6
Sahnow, Will, 191-195, 199, 201, *204*, 206, **288**
'St James Infirmary' 6
St Margaret's, Canning Town (school) 24, 27, 31
St Margaret's, Canning Town (church) 24
Salford College of Technology, aka Salford University, xv, xxxii
Salt, Sir Titus 70, 93-95, **289**
'Salty Dog' xxix
'Salute to Life' 192
Sandmeier, Eric 134
Save Spain From Fascism 57, 143
Save the Rosenbergs 143
Scandinavia Mills 105
Schoenberg, Arnold 188
Schofield, Derek xxvi
Schnabel, Artur 64
Scotland Sings 195
Scott's Motorbikes 97
Scrooge 173
Seal, Peter *289*, **289**
Search For Five Finger Frank, The 123
Searle, Ronald 169
Sebastian, Victor 7
Second Viennese School 68, 188
Seeger, Peggy 222, 233
Seeger, Pete 32, 211, 217
Seiber, Mátyás 197-198
Shadow of a Doubt 177
Shakespeare, William 175, 177

Sharp, Cecil xxii-xxiii
Sharp, Gerry 206, 209, 212, 229-231, *290*, **290**
Shaw, George Bernard 63
Shaw, Susan 169
'She Moved Through the Fair' 197
Shipley Guardian 100
Shipley Youth Club 97-98, *100*, 116
Shostakovich, Dimitri 192
Shuel, Sally 60, *62*, 169-171, 232-233, *291*, **291**
'Shule Agra' 197
Shuttle and Cage 195, 232
Sibelius, Jean 117
Sidebotham, Ian xxix
Sidelman, Ned 201, 212
Siegmeister, Elie 194
Simpson, Martin xiv
Sims, Hylda 87-88
Sinclair, Upton 30-31
Sing 99, 123, 142, 145, 148, 185-186, 217-218
Sing As We Go 159
Sing Out 185-186, 209
Singing Englishman, The 194
Singing Sailor, The 218
'Six Jolly Wee Miners' 123
'Slap Dab' 8
Slater, John 162, 165
Sloane, E. 136
Smart and Brown (Engineers) Ltd 80-83
Smee, née Leader, Patricia (b.1938) *13*
Smiles of a Summer Night 183
Smith, Albert 77
Smith, Donald 106, *107*
Smith, Harry 213
Snagge, John 54
Songs From the Hebrides xxii
Sounds of North American Frogs 207
South Hallsville bombing 78-79
South Riding (book) 25
South Riding (film) 108
'Soviet Airman's Song' 194
Sowerby Bridge Station 109
Spanish Civil War 53, 57, 59-60
Spector, Phil xv
Spencer Watson, Hilda and Mary 187
Stage Fright 173
Stalin, Joseph xxxi, 44, 53, 143-145, 180-181
Stephens, Robert 116
Stevenson, R.L. 190
Stewart, James 183
Stinson Records 207-208
Stokowski, Leopold 160
Stoppard, Tom 175
Storm Over Asia 172
St Pancras Town Hall 117
Strada, La 183
Stravinsky, Igor 142, 186-187
'Streets of Laredo' 6
stripeys 232-234
Sugden, Arnold 103-104, 222
Sullivan, Arthur 23
Summerhill school 88
'Sweet Inniscarra' xxix
Swettenham, Dick 199, 220, **292**
Swinging to the Left 67
Swingler, Randall 63-64
Symphony no. 2 (Sibelius) 117

Tafler, Sidney 169
'Talking Rearmament' 138
Tati, Jacques 157
Taylor, Elizabeth (author) 178
Taylor, Joseph xvii-xviii
Tchaikovsky, Pyotr Ilyich 163-164
Tearing Tickets Twice Nightly 7, 226
'Tell Me What is True Love' xxix
Terry, Sonny 210-212
Tetley, Joyce *100*, 116
Thayer, Webster (judge) 35
Theosophical Society 26
There Will Be Blood 30
'They Don't Write 'Em Like That Anymore' xxxiii-xxxiv
Thief of Baghdad, The 173
Third Man, The 177
Thistlethwaite, Frank 102-104, **293**
Thomas, Dylan 175

Thompson, E.P. 117, 140
Three Score and Ten xv, 194
Titanic 28
Todd, Anne 167
Tom and Jerry xxxiv
'Tom Dooley' 199
Topic Folk Club xxxii,, 157
Topic Records xiv-xv, xvii, 63, 141, 197, 206, 215-217, 229, 233
economics, 192-194, 215-220, 224
and Folkways, 208-210, 212
formats and prices 216-219, 221-222
and Riverside 208
and Stinson, 207-208
survival strategy 231-232
Topic Record Club 191-194, 217
Toscanini, Arturo 164
Tovar González, José Luis *58*, 59-60, **293**
Townshend, Peter (royal suitor) 180
Traditional Folk Songs of Japan 213
Trial by Jury 23
Trailer (label) xiii-xv, xvii, xxxii, 184, 213
Tranco 220
Trains and Buttered Toast 63
Transatlantic Records xiv, xvi, xxxii
Travelling Folk: Leader's Tapes xxxiv
Trewby, G.C. 4
TRICO, aka The Rustless Iron Company 81, *84*
Trotsky, Leon 7
Trouble With Harry, The 176-177
Tunney, Paddy xvi-xvii
Türkbas, Özel 207
Turner, Bruce 205
Twenty Soviet Composers 194
Twisted Nerve 189

Ulysses (book) 169
Unity Theatre 63-68, 217, 226, 321
Ustinov, Peter 208

van Gyseghem, André 67, **294**
Vanzetti, Bartolomeo xviii, 33, 35, 37
Vari-Typer, the 195-196, 209
VAT 223-224
Vaterland, SS 29
Vaughan Williams, Ralph 118, **294**
Vaughan Williams, Shelley 119
Veidt, Conrad 166
Vickers Armstrong 57, 59, 74
Victoria and Albert Museum, the 68
Victoria Hall, Saltaire, 94, 127
Violin Concerto in A Minor (J.S. Bach) 117
Voice of the People, The xviii
Vox Pop, see *Bulletin, The*
Vyse, John 197-198

Wajda, Andrzej 185
Walbrook, Anton 168
Waldron, Bridget (1870-1944) 2, 11, *50*
Wallis, Heather 119-120, *120*, 148
Wanted For Murder 167
Warner, Jack 76, 169
'Warsaw Ghetto Song' 198
Wasson, R. Gordon 209
Watersons, the xiv
Watson, Bob xxix
Watson, Rev. John Selby 96
Waugh, Evelyn 99
Weavers, the 199
We Dive at Dawn 114, 116, 167
Weil, Kurt 64
Welch, Chris 200
Welles, Orson 174-175
Wells, H.G. 63
Went the Day Well? 165
West, Don xxxiii
West, Hedy xxxiii
Wharfedale Wireless Works 103
Wharncliffe, Earl of (first) 154
Wharncliffe, Earl of (fourth) 154, 156-157, 205
Wharncliffe, Lady Montagu 154
When Gazelles Leap 119
Where Justice Faltered 96
Whiskey, Nancy 201
Whitaker, Malachi 112
White, Mary Elizabeth (Molly) (1898-?) 19, 24-25, 44, 46

White, Mary Louise, see Jones, Mary
Whitelaw, Billie 116
Whiteman, Mr and Mrs 66
Whittington, Edith *124*, 125
Whittington, Llewelyn *124*, 125
Whittington, Pauline 125
Whittington, Tony 122, *124*
'Why Do Your Bob Your Hair, Girls?' 166
Wicks, William Francis (Grandad) 79-80
Wilde, Brian 166
Wilde, Oscar xxv
William - The Detective 70
Williams, Bransby 163
Williams, Elin 144
Wiliams, Kenneth, *152*
Williams, Lenin 154
Williams, Vin 154-155, 157
Wills, Charlie xvii, xxii
'Will the Circle be Unbroken' xxix
'Wilno Ghetto Song' 198
Wilson, Woodrow 29
Wilton, Robb xxvii
Windhill Industrial Co-operative Society 97
Winnington, Richard 158
Winter, Audrey xxx-xxxi, **295**
her apostasy, 143-145, 182
Winter, Eric xxx-xxxi, 99, 142-145, 185, *295*, **296**
Winters, Joan 76
Winter, Sue 143
Withers, Googie 169
WMA Singers (Bradford) 153
WMA Singers (London) 192, 198
WMA Summer School 154, 185
Wodehouse, P.G. 200
Woodacre, Curly *152*
Woodbine and Ivy Band xxii
Wood, Peter 175-176
Woods, S. John 169
Woolworths 221
Workers' Education Association 117, 140
Workers' Music Association xiv, xvii, xxiii, xxxii, 63-65, 67, 141, 153, 164, 186, 188, 191-195, 206, 208, 215-217
Annual Report 193-194, 201, 216
Archive, 196, 211
publications 194-195
Summer School, 201-202
Working Class Movement Library 65, 162
World Cinema 183-184
World Federation of Democratic Youth 185
World Festival of Youth and Students 185
Wortley Hall 128-129, 154-157, 202, *203-204*
Wortley, Rick 157
Wright, Harry 134
Wright, Willie *124* (and wife, *124*), 125, 138, *139*

YCL Choir (Leeds) 134
Yorkshire Symphony Orchestra 119
Young Communist League (YCL) xi, 113-114, 129, 134
Young and Innocent 177

The following volumes in *SOUNDING THE CENTURY*, each to have illustrations by PETER SEAL and rare pictures, are in an advanced state of preparation –

VOL. 2. HORIZONS FOR SOME: 1956-1963

Enter Ewan MacColl – enter Bert Lloyd – married life with Gloria – John Hasted and the skiffle boom – the City Ramblers in Germany and Denmark – enter Shirley Collins – Bill and the Brooklyn Cowboy (Ramblin' Jack Elliott) – the US influx inc. Tom Paley, John Gibbon, Peggy Seeger, Guy Carawan – Paul Robeson and the first transglobal concert – London's extraordinary John Foreman, the Broadsheet King –the strange story of Aubrey Pankey – the equally strange yet in some ways similar story of Perry Friedman – building a studio in Bishops Bridge Road – anti-nuclear protest in England and Scotland (*sing like Eskimos!*)

VOL. 3. THE POOR MAN'S ONLY MUSIC: 1957-1963

Mostly source singers and field-recording do's and don'ts – Joseph Taylor – Walter Pardon – Jeannie Robertson – Cait O'Sullivan, a grievous Lost Leader – Bill and the Coppers – Reg Hall and Bob Davenport, swallows in springtime – the world of Deben Bhattacharya

VOL. 4. BAD LADS AND HARD CASES: 1959-1966

Enter Nat Joseph – birth of Transatlantic – the Australian connection (Barry Humphries, Peter Mann) – Felix Raab, a tragic tale – the underworld element and inglorious behaviour – The Black Horse Broadside – Dominic Behan – a boom in clubs and EPs (linked phenomena) – Bill and the Beatles – Bill in Edinburgh – the sexual revolution and marriage break-up – enter the Watersons – Bert Lloyd and 'Bert songs'

VOL. 5. LET US NOW PRAISE OBSCURE MEN: 1962-1967

In praise of obscurity – Sandy and Jeanie come and stay – Paul Simon, a Lost Leader – Pink Floyd, a Lost Leader – Hedy West – Anne Briggs – enter Bert Jansch – Peter Bellamy – the Campbells – marriage to Helen – a field-trip to Ireland...

www.soundingthecentury.com

Matador